Advanced Microwave Circuits and Systems

Advanced Microwave Circuits and Systems

Editor

Lefteris Tyler

scitus academics

Advanced Microwave Circuits and Systems
Edited by **Lefteris Tyler**

ISBN: 978-1-68117-209-5
Library of Congress Control Number: 2016934756

SCITUS Academics LLC,
www.scitusacademics.com
Box No. 4766, 616 Corporate Way,
Suite 2, Valley Cottage,
NY 10989

Printed in United States of America on Acid Free Paper

Preface

Microwave systems, such as radars, and radio and other communication devices are made of many electronic parts called "circuits" or "components." These are designed and constructed to manipulate electromagnetic phenomena into carrying out different microwave signal processing functions such as generating, modulating, controlling, and amplifying signals. They are also used in frequency translation (conversion). The construction of such electronic circuits is usually quite different from those used in low-frequency equipments, such as television sets or AM/FM radio. In the early development of microwave circuits and systems (during and just after World War II) heavy and bulky microwave circuits in the form of voluminous, hollow metallic pipes and tubes were used. These large, three-dimensional waveguides are still in use today for certain high-power applications.Microwave engineering pertains to the study and design of microwave circuits, components, and systems. Fundamental principles are applied to analysis, design and measurement techniques in this field. The short wavelengths involved distinguish this discipline from Electronic engineering. This is because there are different interactions with circuits, transmissions and propagation characteristics at microwave frequencies. Some theories and devices that pertain to this field are antennas, radar, transmission lines, space based systems (remote sensing), measurements, microwave radiation hazards and safety measures. The book, Advanced Microwave Circuits and Systems, deals with the design and development of active and passive microwave components, integrated circuits and systems. The book accomplishes with chapters considering solicitation of microwaves in measurement and sensing systems.

Table of Contents

Chapter 1

Modern Monitoring Intraocular Pressure Sensing Devices Based on Application Specific Integrated Circuits

Daniel Piso[1], Patricia Veiga-Crespo[2], Elena Vecino[2*]

[1]Group of Research in Control and Diagnostics, ESS Bilbao, Vizcaya, Spain;
[2]Group of Experimental Ophthalmo-Biology, Department of Cell Biology and Histology, University of the Basque Country UPV/EHU, Vizcaya, Spain.

ABSTRACT

Glaucoma is a neurodegenerative condition that is the leading cause of irreversible blindness worldwide. Elevated intraocular pressure (IOP) is the main risk factor for the development and progression of the disease. Methods to lower IOP remain the first line treatments for the condition. Current methods of IOP measurement do not permit temporary noninvasive monitoring 24-hour IOP on a periodic basis. Ongoing research will in time provide a means of developing a device that will enable continuous or temporary monitoring of IOP. At present a device suitable for clinical use is not yet available. This review contains a description of different devices currently in development for measuring IOP: soft contact lens, LC resonant circuits and on-chip sensing devices. All of them use application-specific integrated circuits (ASICS) to process the measured signals and send them to recording devices. Soft contact lens devices are based on an embedded strain gauge, LC circuits vary their resonance frequency depending on the intraocular pressure (IOP) and, finally, on-chip sensing devices include an integrated microelectromechanical sensor (MEMS). MEMS are capacitors whose capacity varies with IOP. These devices allow for an accurate IOP measurement (up to

+/– 0.2 mm Hg) with high sampling rates (up to 1 sample/min) and storing 1 week of raw data. All of them operate in an autonomous way and even some of them are energetically independent.

Keywords: Glaucoma; Intraocular Pressure; Monitoring; Implantable Device; Application-Specific Integrated Circuit

1. INTRODUCTION

Glaucoma is the second cause of blindness. It affected to approximately 60 million people worldwide in 2010 and there are predictions of noticeable increases within the next decade [1]. It is a progressive, asymptomatic disease of the optic nerve. In the normal eye, aqueous humor is produced in the ciliary body in the anterior chamber at balanced rate. Intraocular pressure (IOP) is the balance between aqueous humor formation and outflow. It circulates throughout the eye and finally drains through the trabecular meshwork into the Schelm's canal and the episcleral veins [2].

Failure of the aqueous humor to leave the eye due to the obstruction in the drainage sites result in elevated IOP. There are different reasons: an increase in the epiescleras venous pressure [3] or an increase in trabecular resistance outflow [4]. Elevated IOP is not the only diagnostic factor and many people with high intraocular pressures have normal vision. However, 1 mm Hg increase in IOP increases the risk of getting glaucoma by 10%.

IOP varies widely through the day (ranging from 12 - 21 mm Hg) and it is even pulsated according to its cardiovascular origin. Therefore, two or four measurements obtained periodically during the year are not indicative of patients actual IOP. Current methods of IOP measurement do not permit temporary noninvasive monitoring 24-hour IOP on a periodic basis. Such measurements may be critical in understanding the progression of glaucomatous visual loss, especially in normotensive or lowtension glaucoma. Despite numerous previous attempts at continuous IOP monitoring, a device suitable for clinical use is not yet available. However, devices currently in development for permanent IOP monitoring seem to be nearly ready for human testing.

IOP is known to undergo both random fluctuations as well as variations following a circadian pattern. In humans, IOP is highest at night and lower during the daytime, largely due to changes in body position, although other factors appear to contribute. Ongoing research will in time provide a means of developing a device that will enable continuous monitoring of IOP and should lead to improved management of glaucoma with attendant benefits in reducing glaucoma blindness.Thus, there is a need for cost-effective measurement devices for early management of glaucoma.

Classically, the two most widely employed devices to measure IOP were the Goldmann tonometer and the Tono-Pen, which is a microprocessor-controlled portable version of the Goldman tonometer [5]. Both devices utilise the applanation principle, which is based on the relationship between force, pressure and area.

Although the Goldmann method is considered the standard procedure for measuring IOP, it has a number of disadvantages. For the patient, it is highly invasive and uncomfortable. First, the head of the tonometer must come into direct contact with the cornea. Second, in order to prevent reflexes during measurement, anaesthesia must be applied to the numb eye. A fluorescein strip must also be applied to the eye to detect the applanated state. All these steps require a skilled technician or expertise of the ophthalmologist. And, despite the advanced techniques involved, they do not provide an accurate measure of the patients IOP.

The increasing capabilities of the electronics industry for sensing signal processing, actuating and communication with the environment have been applied to health care for years. One of the most spectacular topics in this field is implantable devices. This industry is not completely new. From battery-powered pacemakers, semiconductor electronics have been used to devise implantable medical devices that save lives and improve patient quality of life.

During the last years, the implant volume and other characteristics have been dramatically improved by orders of magnitude [6]. This way, this technology was transferred into new therapy areas such us the measurement of glucose levels, devices interfacing with the body's electrical network or breath neuromodulation technologies.

Additionally, smaller devices with better performance increases patient acceptance for therapies targeting prevention and enable less invasive and lower cost procedures. Moreover, the advanced miniaturization allows the using of new form factors and even independent devices that communicate with each other as a part of a Body Area Network.

One of the last cutting-edge working areas for implantable devices is their application to monitor glaucoma intraocular pressure (IOP) [7]. So, the former theoretical idea of an autonomous intraocular implantable device using Application Specific Integrated Circuits (ASICs) is very close to become a reality. This paper is a review consisting of an analysis of the main principles that new devices rely on and some of the more new, advanced and representative devices.

Currently some different ideas for implantable devices are being developed. One of the techniques is the integration of a strain gauge pressure sensor into a soft contact lens [8]. This system was initially wired and then made wireless. These devices produce measurements affected by variability, which has to do with the possible movement of the lens. There are other implantable devices that do not suffer from this problem. They are based on LC resonant systems [9]. This solution produces robust, simple and smaller solutions. Conversely they have limited functionality.

Along the past two years, the work of some research groups in implantable devices has produced an important leap in the technology for continuous IOP monitoring [10,11]. A set of new devices consisting of ASICs and microelectromechanical systems (MEMS) has been published. This new systems provide operation over a much longer monitoring period. Recording and measures are done up to every 5 minutes. For instance, one of them consists of an ASIC fabricated with a 0.13 μm technology, MEMS, antenna and capacitive power array. The MEMS sensor is used to convert pressures into capacitance and is fed into the ASIC for data processing and modulation onto a high-frequency carrier for wireless transmission.

In the present review, we will describe the classical Goldmann based measuring methods (tonometry). Then, the main new methods based on Application Specific Integrated Circuits and their theoretical basis and finally, some of the most representative methods of each category are analysed. This review ends with some concludeing remarks.

2. CLASSICAL IOP MEASUREMENT METHODS

The main different principles developed in the past to measure the IOP are: palpation, manometry and tonometry.

Palpation: is also known as digital tonometry. It estimates IOP by gently pressing the index finger against the cornea of the closed eye. This method is notoriously unreliable and the oldest one. It is the simplest, cheapest method but also the least accurate method to estimate the IOP [12]. By comparing tactile estimations of IOP to formal pressure measurements, the examiner's sense of touch can be educated up to a limit. Bowman suggested in 1862 [13] a classification of the examined tension divided into 9 grades (T1 to T3 and –T1 to –T3), which ranges from "extremely low tension" to "maximally raised tension". This classification was the first method for quantitative estimation of pressure differences.

Manometry: This is an invasive technique that measures IOP in a more accurate way. It is used mainly as a laboratory technique to perform continuous IOP measurements for the evaluation of experimental pharmacological preparations [14]. Ethical rules indicate the using of this method only for eyes undergoing enucleation or intraocular surgery [15].

Tonometry: It is the process of determination of IOP by means of a device called tonometer. This device determines a value for IOP with minimum damage for the patient's eye. It is based on the Pascal principle. The pressure applied to an enclosed fluid is transmitted undiminished to every part of the fluid, as well as to the walls of the container. So the working principle consists of trying to measure the IOP accessing to the cornea. This is the only ocular structure, which is externally accessible. Depending on the particular physical principle applied (indentation, applanation, contour matching, rebound), we found different types of tonometers. Any device uses one or a combination of these principles.

2.1. Applanation Tonometry

In this case, the IOP is inferred from the force required to flatten a constant area in the cornea (**Figure 1**). The physical principle is the Imbert-Fick law [16]: "the pressure in a sphere filled with liquid and surrounded by an infinitely thin membrane is measured by the counter pressure which just flattens the membrane". The parameters of this law are related in the following way:

$$\mathrm{IPO} = \frac{\text{contact force}}{\text{contact area}}$$

According to this expression, there are two ways of determining the IOP: measuring the force necessary to flatten a fixed area (fixed-area tonometer) or measuring the area flattened by a fixed force (fixed-force tonometer). This law assumes an infinitely thin, perfectly elastic and flexible cornea. So the thickness and the resistance offered by the cornea when is pressed are sources of error [17].

2.1.1. The Maklakoff Tonometer

This is a fixed-force tonometer and an early example of this method. This device was developed by von Graefe and Maklakoff in the XIX century. Due to direct contact with the eye, anaesthesia must be applied in the eye as eye drops. IOP was significantly underestimated compared to more modern methods. Other example of this type of devices is the Posner tonometer developed in 1965.

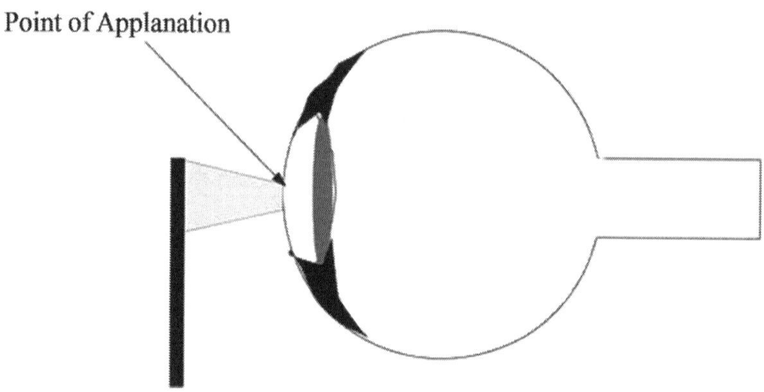

Figure 1. Applanation tonometry.

2.1.2. The Goldmann Tonometer

This is the most widely used tonometer, which is based in the fixed-area principle. The applanation surface has a diameter of 3.06 mm placed in the centre of a plastic cylinder. This cylinder is attached to an arm pushed forward through a spring-loaded knob. The amount of force on the cylinder is precisely controlled and can be read from a scale on the knob. The device is mounted on a slit-lamp biomicroscope. In this case, topical anaesthesia is required together with a fluorescein dye. When this tonometer is employed some variables can affect on the IOP measurements: corneal rigidity can overestimate the IOP; tears can underestimate the measurement and high corneal astigmatism can cause big deviations.

This device is considered to be the gold standard test and the most widely accepted method. However, it is not portable. Perkins tonometer was designed with the same principles but it is a portable one, battery powered and usable in the supine as well as sitting positions. One of the main disadvantages of contact tonometers is the spread of infections if the devices are not adequately sterilized between measurements.

2.2. Dynamic Contour Tonometry

Dynamic contour tonometry (DCT) uses the physical principle of contour matching instead of applanation. The DCT technology is a noninvasive and direct IOP measurement. This technology has been proposed to continuously and accurately measure the IOP of structural characteristics of the cornea and sclera [18].

A practical implementation of DCT tonometry is the PASCAL tonometer. The PASCAL tonometers have a tip that contains a hollow with the same shape as the cornea and a miniature pressure sensor in its centre. In contrast to applanation tonometry, it is designed to avoid deforming the cornea during measurement. Therefore, thought to be less influenced by corneal thickness and other biomechanical properties of the cornea, it is more influenced by corneal curvature. The PASCAL IOP is defined as the diastolic IOP.

2.3. Indentation Tonometry

The basis of this tonometry is: a known force will indent a fluid or gas filled object to a greater degree if the internal pressure is low compared with when the internal pressure is high. Van Graefe, in 1862, was the first that tried to apply this physical principle in a tonometer. In 1905, Schiotz developed the Schiotz tonometer that soon replaced the Maklakoff tonometer. Other indentation tonometers were designed later such as Bailliart tonometer or Maurice Electrical tonometer but only the Schiotz's is still in clinical use today [19]. The MacKay-Marg tonometer, the TonoPen tonometer and the pneumatonometer device have properties of both applanation and indentation. The original MacKay-Marg tonometer uses a microplunger that protrudes a small amount from a flat footplate of a tubular hand piece. The microplunger is connected to a sensitive transducer, which converts pungler displacement into an electrical signal that is recorded on a paper chart, much like an electrocardiogram [20]. This device has been totally replaced by battery-operated and handheld version, the TonoPen.

In the TonoPen, the tip is covered by a disposable latex cover and applied perpendicularly to gently indent an anesthetized cornea. Each measurement requires several applanations. An acceptable applanation is indicated by an audible click after contact with the cornea. A microprocessor averages several acceptable waveforms and gives a digital readout of IOP on a liquid crystal display, with an estimate of the variability between the component readings. The TonoPen tends to give higher reading than the Goldmann applanation tonometer. Finally, the Pneumatic tonometer allows a continuous IOP reading that is recording on a moving paper chart.

2.4. Rebound Tonometry

Rebound tonometer determines intraocular pressure by bouncing a small plastic tipped metal probe against the cornea. This principle was applied by Obbink by first time in tonometry in 1931. The most recent version is the iCare tonometer. The iCare tonometer was introduced in 1997 [21]. The handheld device uses a projectile of stainless steel. One end is covered with a small plastic cap with a radius of 0.9 mm. Considering all of this, the ideal tonometer would be easy to use, comfortable for the patient and the physician, portable and reliable. A version of this tonometer for the using in small experimental animals has become very popular and one of the most used in research.

3. NEW IOP MEASUREMENT TECHNIQUES

There are different possibilities to classify the new IOP measurement techniques. In this paper, a classification depending on the used technology is proposed. There are three main groups: those that are based on soft contact lens, based on LC resonance and on-chip sensing devices. The last two groups produce implantable devices (placed in the anterior chamber) and show some advantages for pressure measurement, as they are independent of ocular surface, cornea rigidity, etc.

3.1. Wireless Contact Lens Sensors Devices

This noninvasive pressure sensing technique is based on a strain gauge embedded in the lens (see**Figure 2**). This lens is located over the meridional angle of the corneoscleral junction. The sclera is the preferable site for the sensor because of lower sensitivity and less susceptibility to injury than cornea [22].

The strain gauge is able to measure the angular changes at the junction due to variations in IOP. In order to work properly, the lens must be moulded as an exact copy of the eye surface.

The primary designs of this technique used wire lead from the strain gauge [23]. They were passed through the holder and over lower eyelid. These wires were connected to a recording device. However, later refinements of this technique made the recording process wireless, as it is explained later. Leonardi et al. [24], which is a representative design, built a design based on a micro-fabricated strain gauge in Wheastone bridge configuration.

The changes in corneal curvature correlated to intraocular pressure are measured. In general, there is a good correlation between the real pressure and the results obtained from the sensor. The device is placed in the same way as a corrective contact lens, no anaesthesia is required and vision remains unaffected.

The strain gauge mentioned before is inserted in a Wheastone bridge configuration [23], with two sensing resistive gauges (double sensitivity), and two compensation resistive gauges (thermal compensation). Under these conditions, a soft contact lens bends and follows corneal deformations, behaving like a partial spherical shell subjected to a uniform edge moment and has only circumferential strain. The sensing resistive gauges in the device are designed to have a circular arc shape around the centre. They have 11.5 mm of diameter, the same as the average corneoscleral junction position. It is supposed that at this place, the IOP changes cause the maximum corneal deformation.

The sensor contained in the lens is made of a thin microfabricated platinum (170 nm Pt/25 nm Ti). The gauge is embedded between two layers of polyimide, which is suitable for biomedical applications and offers advantages compared to other materials. The sensor is positioned in the mould and the outer shape of the lens is polymerized (**Figure 3**). Then silicone is added and polymerized to form the inner shape.

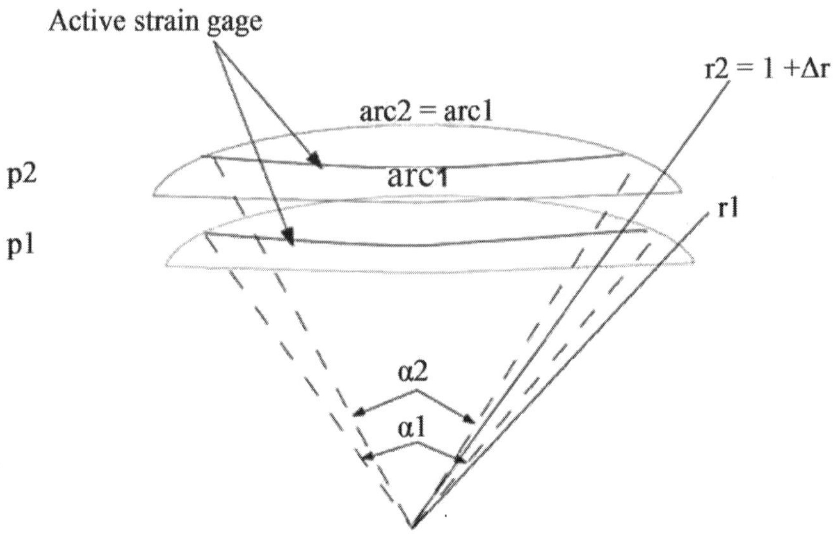

Figure 2. Deformation of a soft contact lens when IOP changes from pressure p1 to p2.

Initial designs included a 3 cm microflex connection cable, as well as a contact area for a zero insertion force connection [22,23]. However, the using of wires is a major drawback of the system: discomfort for the patient and air bubbles between the eye and the cornea. So, subsequent improvements of these devices resolve the powering and communication between the contact lens and an external recording unit performing it wirelessly by a microprocessor and an antenna integrated into the soft contact lens. In [24], the design of a dedicated application-specific integrated circuit (ASIC) for wireless communication is described. This ASIC has a thickness of 50 µm, it is connected to a sensor and a 10 mm diameter gold antenna was integrated. The final thickness of the lens is about 400 µm at the centre and 100 µm at the border.

The embedded ASIC transmits an output signal V_m proportional to the contact lens strain and therefore the IOP. Data are stored in a portable unit, worn by the patient. The measurement principle is based on stimulating the Wheastone bridge via a DC current I_0. The bridge output is a signal voltage V_m proportional to its strain.

$$V_m = \frac{\varepsilon \cdot GF \cdot I_0 \cdot R_0}{2}$$

where ε is the strain of the active gages, GF is a gage factor and r the initial resistance of the gages. The lens should follow the corneal deformations, so ε can be written in terms of the initial curvature of the contact lens rand the initial opening angle α. The relation between ε and the variation of the curvature (Δr) is not linear. However, as these variations can be considered very small it is possible to find a simple and linear expression that contains the main idea of this approach: the corneal curvature is related to the IOP and, hence to the obtained voltage V_m.

$$V_m = \frac{\alpha \cdot \Delta r \cdot GF \cdot I_0 \cdot R_0}{2}$$

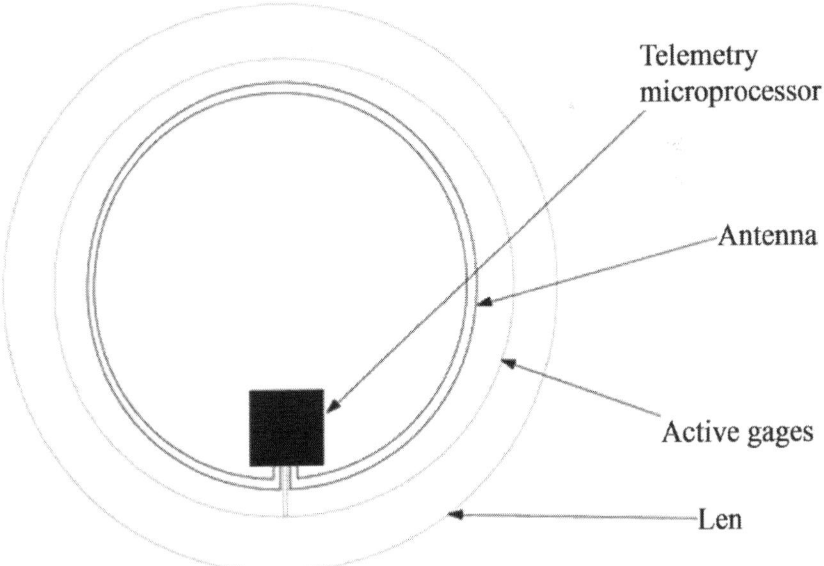

Figure 3. Contact lens wireless sensor.

Finally, the change in the cornea curvature is related to a pressure variation in the following way.

$$\Delta p = \frac{2 \cdot b \cdot V_m}{\alpha \cdot GF \cdot I_0 \cdot R_0}$$

The test bench designed for this device is based on measuring the IOP in enucleated pig eyes with the contact lens sensor. An eye was cannulated with a needle inserted into the anterior chamber and connected with a silicone tube to a column filled with saline, whose amount is controlled by a syringe pump. A pressure sensor in the silicone tube permits the measurement of the pressure real value. The pressure from the contact lens was compared to this one. The contact lens sensor is powered wirelessly by the portable unit through a loop antenna and through the same wireless connection. The obtained data were recorded in a portable unit and sent to a PC.

This CLS showed a high linearity (R^2 = 0.9935), a sensitivity of 109 µV/mm Hg and a reproducibility of +/– 0.2 mm Hg. So it is demonstrated that the device would be able to measure a signal equivalent to the human ocular pulsation and follows the IOP in a reproducible way.

3.2. Inductive Couple Telemetry LC Resonance Based Device

Almost all of them use capacitive transducers for pressure sensing [25]. The most common system is a passive LC resonance whose resonance frequency changes with pressure [26-28]. A coil is bonded to the capacitive pressure transducer element. As the capacitive coupling inside the coil changes, the resonance frequency of the whole system changes with pressure. Some devices, must be used to remotely sense the pressure using this technique.

These devices suffer from the requirement of high coupling between the external antenna and implant. So the optimal performance occurs when the sensor and detection circuits can be close to each other.

Traditionally, it was assumed that LC resonant devices were less accurate and with less sensitivity. However, the advances in materials and micromachining have improved the performance of this technique.

In [30] an IOP measuring device based on an inductance-capacitance (LC) resonant circuit is presented. As it was explained before, the capacitor in the LC circuit is used as pressure sensing device. Hence, the capacitance of the sensing device is related to the IOP. When the IOP changes, the capacitance in the circuit is modified and therefore, the resonance frequency also changes. So the measurement principle consists of measuring the resonance frequency of the circuit.

An external coil, placed near the eye, generates a magnetic field whose field lines are perturbed by the sensor. This produce changes in the impedance of the external coil ($R_L(w)$) as a consequence of the variations of the IOP. Then the measure is made by extracting the real part of the impedance of the external coil as a function of the frequency, which depends on the quality factor of the LC circuit itself and the coupling factor between the two coils (this quality factor gives the measurement resolution). Finally, the voltage measure of the external coil gives the impedance and then IOP.

$$V_{out} = \frac{A_0^2 R_L(\omega)}{2}$$

The system is made up of two differentiated parts: the analog section (in charge of the measuring) and a microcontroller. The ASIC microcontroller does not need to be high-speed processing element because the major part of the processing is made in the analog part. The analog/digital conversion and the frequency meter are contained in the microprocessor. The data are stored in a 256 Kilobytes memory. The IOP is recorded at a rate of 1 sample per minute. It is able to store 1 week of raw data that can be uploaded to a personal computer afterwards.

The passive sensor is implanted between the eyeball and the orbital bones where it can be fixed by means of a simple surgical access.

3.3. Capacitive On-Chip Sensor Devices

This is the most modern and technology developed approach. It consists of a programmable intraocular pressure sensor system implant integrated on a single CMOS chip. It contains on-chip micromechanical pressure sensor (MEMS) array, a temperature sensor, an antenna, a capacitive powering array, readout and calibration electronics, a µC-based digital control unit, and an RFtransponder. These devices are more robust for miniature implants and they permit to transmit the power and data to longer distance. On-chip storage of calibration data, high signal-to-noise ratio are other advantages. RF fields are usually used to activate and power sensor chips without using a battery [29]. RF is also used to transmit the data digitally to a remote reader unit. The pressure sensor consists of a pressure dependent capacitor (MEMS) used to convert pressures into capacitance. This information is processed by the ASIC and then, modulated into a highfrequency carrier for wireless transmission [10,11].

These types of sensors improve the possibilities for glaucoma disease treatment. There is a great chance for increasing the devices functionalities. In addition to this, it allows for the measurement for extended periods of time without having the presence of an external device providing power or extracting data.

Measurement principle One of the most representative example of these devices is described in [10,31]. The ASIC is constructed by using a 130 nm technology and is made of four parts: a sensor interface, voltage regulators and references, radio-frequency rectifier for remote powering and a wireless transmitter. An integrated monopole antenna is in charge of the wireless transmission and the independent operation is guaranteed by a capacitor array. That fits within the size constraints of the miniature package to provide power storage and avoid an external power supply.

The measurement of the IOP is based on two capacitors (see **Figure 4**). On of them is a MEMS, whose capacitance varies depending on the pressure. Therefore, whenever the IOP varies, so does the MEMS capacitance and, hence its charging/discharging time. So measuring the charging time it is possible to derive the IOP value. The other capacitor (Base cap in **Figure 4**) acts as a reference and its capacitance is equal to the base capacitance of the MEMS.

The MEMS sensor always has an equal or larger capacitance, compared to the reference capacitor, resulting in a longer charge-up time. Digital logic is used to create a pulse transition whenever one of the two capacitors charges up the corresponding Schmitt trigger in **Figure 4**. The voltage in the MEMS capacitor is:

$$V_{MEMS} = t \cdot \frac{I_D}{C_{MEMS}}$$

t is the time while the a current is present, C_{MEMS} is the capacitance of the MEMS sensor and I_D is the current supplied to the capacitors. The initial current values are greater to produce a fast charging up to the base capacitance of the sensor. After this, the remaining capacitance, which represents the pressure reading, is charged with a smaller current value. This way, larger time variations are needed for this process and the sensitivity is increased. This effect is crucial as we are considering a low-power, low-delay device. The final time expressions is

$$t = \frac{V_{MEMS} \cdot C_{MEMS}}{k \cdot \left(1 - \lambda V_{DD} - \lambda V_{MEMS}\right)}$$

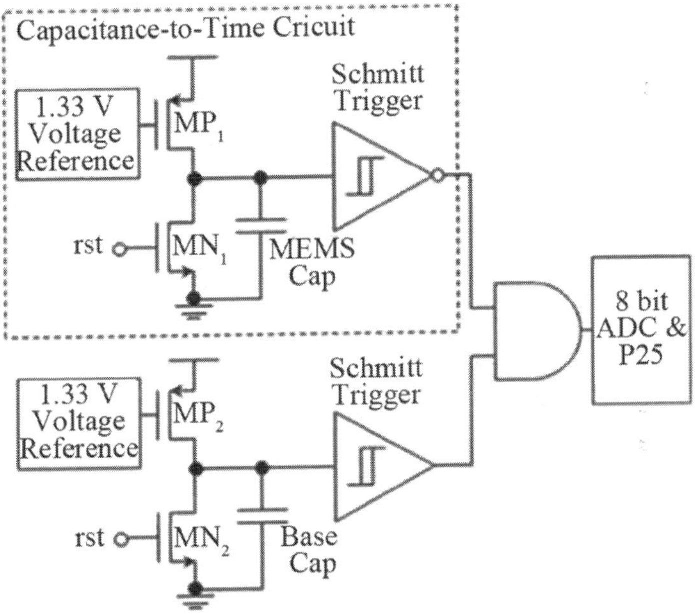

Figure 4. Capacitance-to-time circuit.

V_{MEMS} reaches the threshold of the Schmitt trigger. Then, the output switches from high level to low level. Selecting the value of V_{MEMS} to the threshold in the previous equation it is possible to obtain the time employed to charge the MEMS capacitor. The time between the beginning of the charging up and the threshold is directly proportional to the capacitance of the MEMS and, hence, the IOP. As said before, the reference capacitor reaches the threshold a little bit earlier. This difference represents the MEMS capacitor increment because of the pressure. Therefore, the logic converts this to a pulse, which is further digitized (42.2 KHz) and sent wirelessly or stored in the memory.

To permit the complete independent operation of the device, the system includes a memory storage unit. In this case, a non-volatile memory FeRAM (ferroelectric random-access memory) is used. The wireless data transmission is performed by an antenna (2.4 GHz). This antenna also acts a radio-frequency rectifier for remote powering and recharging. The antenna shape is designed taking into account clinical requirements and electromagnetic optimization (**Figure 5**). Additionally, to satisfy clinical size requirements, a high-dielectric surfacemount capacitors are used to storage the power drawn from the antenna. The device is able to track the IOP in 5 minutes intervals over 24 hours periods with a 0.5 mm Hg resolution.

The final packaging of this device was designed on purpose using a biocompatible hermetic package made of liquid crystal polymer (low moisture absorption, flexible and thermally bonded without adhesives). Biocompatibility tests for this material can be found at [32,33]. The final product has the form of a tadpole, as shown in **Figure 5**. The tail contains the antenna (curved along the rim in the anterior chamber) and the head contain the electronics. The implantation can be made via a trochanter technique. A complementary receiver to get the stored data was also designed Solar energy collecting device.

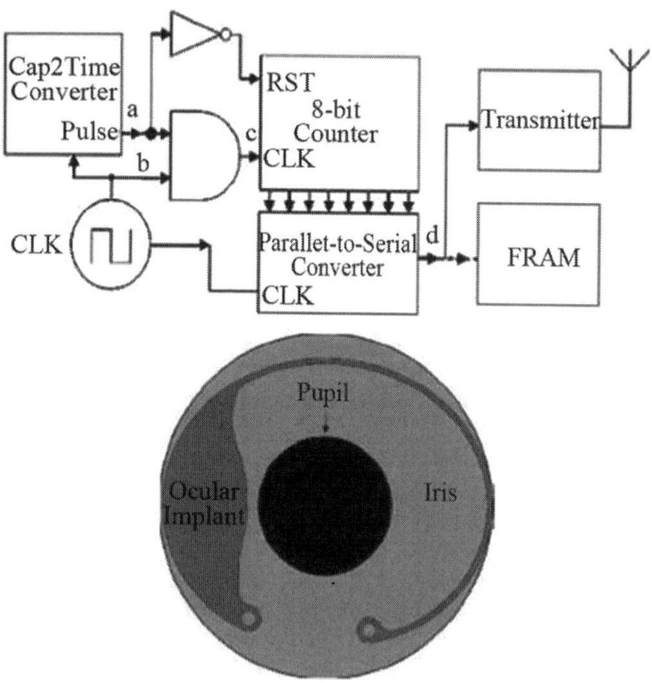

Figure 5. Device block diagram and frontal view of implantation.

An interesting possibility is presented in [11], which consists of a device collecting solar energy instead of getting power from an external device. This energy reaches the eye through the cornea, which is transparent, and it is collected by a 0.7 mm^2 solar cell. This device is made up of an antenna and an electronics part as previously explained.

The energy collection is based on an integrated solar cell, a thin-film Li battery is also present for energy storage. The measurement principle is also based in a MEM capacitive sensor and an integrated circuit. However, in this case the electronics and the antenna are packaged in a compatible glass housing. The main parts of the integrated circuit are the following: a wireless transceiver, capacitance to digital converter, a microcontroller and a memory. The used fabrication technology is 0.18 mm CMOS.

The IOP is measured every 15 minutes with a pressure resolution of 0.5 mm Hg. The measurement technique is very similar to the previous device. A MEMS capacitor and a fixed capacitor are switched at 50 KHz. The fixed current is compared to the IOP-dependent current. The difference between them is processed to produce the pressure measurement. Again, IOP data are stored, in this case in SRAM memory. The microsystem is capable of storing 3 days of raw data. However, an additional functionality is provided that can compress the data to extend the storage capacity to 1 week.

The user can download the IOP data by using an external device, placed near the eye. The microsystem responds to a wireless query by means of an oscillator acting as a carrier generator and an amplifier. The lifetime is limited to 28 days without energy harvesting. The combination of energy harvesting and low-power operation allows the device to achieve zero net energy operation in low light, requiring 10 hours of indoor or 1.5 of outdoor illumination.

4. CONCLUDING REMARKS

This paper reviews the latest different solutions based on application specific circuits for IOP sensing. The main measurement principles used in the different devices have been described: soft contact lens, LC resonant circuits and on-chip capacitive sensor. The last two types generate implantable devices.

Some representative devices of each solution were described. An embedded strain gauge in a soft contact lens permits the accurate measurement by means of an integrated circuit. The sensitivity of this device is +/– 0.2 mm Hg. Wireless operation of these devices has been already devised. The traditional LC resonant circuits have been improved and provide implantable solutions. A device that is able to record 1 sample per minute during 1 week has been presented. Finally, the last cutting-edge devices using a microprocessor with a sensing integrated device have been analyzed. They have an accuracy of 0.5 mm Hg and are able to store big amount of raw data. These devices are able to work wirelessly and independently and have promising improvement possibilities.

These sensing devices are all clear examples of how multidisplinary collaboration between different research fields provides highly promising results. In the future, these devices will be expanded to operate in a closed loop monitoring and treatment system with a great potential for a vast increase in functionality. From a broad perspective, none of these technologies will transform care by themselves, but they can make significant improvements in patients' treatment. In the future, special attention must be paid to reliability, implantation issues, performance and cost. If all these aspects are managed carefully, exciting possibilities will arise, especially to address chronic disease management needs.

5. ACKNOWLEDGEMENTS

Ayudas Grupos Consolidados Gobierno Vasco (IT437- 10), ONCE Foundation (Spain), Red Patología Ocular Inst. Carlos III RETICS (RD07/0062).

REFERENCES

1. H. A. Quigley and A. T. Broman, "The Number of People with Glaucoma Worldwide in 2010 and 2020," British Journal of Ophtalmology, Vol. 90, No. 3, 2006, pp. 262- 267.
2. D. H. Johnson, "Trabecular Meshwork and Uveoscleral Outflow Models," Journal of Glaucoma, Vol. 14, No. 4, 2005, pp. 308-310.
3. T. R. Friberg, G. Sanbor and R. N. Weinreb, "Intraocular and Episcleral Venous Pressure Increase during Inverted Postura," American Journal of Ophtalmology, Vol. 103, 1987, pp. 523-526.
4. N. Weinreb, "Uveoscleral Outflow: The Other Outflow Pathway," Journal of Glaucoma, Vol. 9, No. 5, 2000, pp. 343-345.
5. C. Kniestedt, O. Punjabi, S. Lin and R. L. Stamper, "Tonometry through the Ages," Survey of Ophtalmology, Vol. 53, No. 6, 2008, pp. 568-591.

S. Oesterle, P. Gerrish and P. Cong, "New Interfaces to the Body through Implantable Systems Integration," International Solid-State Circuits Conference, San Francisco, February 2011.

6. K. C. Katuri, S. Asrani and M. K. Ramasubramanian, "Intraocular Pressure Monitoring Sensors," IEEE Sensors Journal, Vol. 8, No. 1, 2008, pp. 12-19.
7. M. Leonardi, P. Leuenberger, D. Bertrand, A. Bertsch and P. Renaud, "First Steps toward a Noninvasive Intraocular Pressure Monitoring with a Sensing Contact Lens," Investigative Ophtalmology and Visual Science, Vol. 45, No. 9, 2004, pp. 3113-3117.
8. C. Po-Jui, D. C. Rodger, S. Saati, M. S. Humayun and T. Yu-Chong, "Microfabricated Implantable Parylene-Based Wireless Passive Intraocular Pressure Sensors," Journal of Microelectromechanic Systems, Vol. 17, 2008, pp. 1342-1351.
9. E. Y. Chow, A. L. Chebowski and P. Irazoqui, "A Miniature-Implantable RF-Wireless Active Glaucoma Intraocular Pressure Monitor," IEEE Transactions on Biomedical Circuits and Systems, Vol. 4, No. 6, 2010, pp. 340- 349.
10. G. Chen, H. Ghaed, R. Haque, M. Wieckowski, Y. Kim, G. Kim, D. Fick, D. Kim, M. Seok, K. Wise, D. Blaaw and D. Sylvester, "A Cubic-Milimeter Energy Autonomous Wireless Intraocular Pressure Monitor," International Solid-State Circuits Conference, San Francisco, February 2011.
11. C. D. Birnbach and M. M. Leen, "Digital Palpation of Intraocular Pressure," Ophthalmic Surgery Lasers, Vol. 29, No. 9, 1998, pp. 754-757.
12. O. Bowman, "On Glaucomatous Affections and Their Treatment by Iridectomy," British Medical Journal, Vol. 2, 1862, pp. 377-382.
13. B. A. Ellingsen and W. M. Grant, "Influence of Intraocular Pressure and Trabeculotomy on Aqueous Outflow in Enucleated Monkey Eyes," Investigative Ophtalmology & Visual Science, Vol. 10, 1971, p. 705.
14. M. Blumenthal, M. Chane and I. Ashkenazi, "Direct intraopative continuos monitoring of intraocular pressure," Ophtalmologic Surgery, Vol. 23, 1992, pp. 132-134.
15. H. Goldmann, "Applanation Tonometry," Transactions Second Glaucoma Conference, Josiah Macy, Jr Foundation, New York, 1957.
16. M. M. Whitacre and R. Stein, "Sources of Error with Use of Goldmann-Type Tonometers," Survey of Ophthalmology, Vol. 38, No. 1, 1993, pp. 1-30.

17. H. E. Kanngieser, C. Kniestedt and Y. C. Robert, "Dynamic Contour Tonometry: Presentation of a New Tonometer," Journal of Glaucoma, Vol. 14, No. 5, 2005, pp. 344-350.
18. D. R. Anderson and W. M. Grant, "Re-Evaluation of the Schiotz Tonometer Calibration," Investigative Ophtalmology & Visual Science, Vol. 9, 1970, pp. 430-446.
19. E. Marg, "A Report on Mackay-Marg Tonometry in Optometry," Journal of the American Optometric Association, Vol. 34, No. 12, 1963, pp. 961-965.
20. A. Kontiola, "A New Electromechanical Method for Measuring Intraocular Pressure," Documenta Ophthalmologica, Vol. 93, No. 3, 1997, pp. 265-276.
21. M. Leonardi, P. Leuenberger, D. Bertrand, A. Bertsch and P. Renaud, "First Steps toward Noninvasive Intraocular Pressure Monitoring with a Sensing Contact Lens," Investigative Ophthalmology & Visual Science, Vol. 45, No. 9, 2004, pp. 3113-3117.
22. M. Leonardi, P. Leuenberg, D. Bertrand, A. Bertsch and P. Renaud, "A Soft Contact Lens with a MEMS Strain Gage Embedded for Intraocular Pressure Monitoring," 12th International Conference on Solid State Sensors, Actuators and Microsystems, Boston, 8-12 June 2003.
23. M. Leonardi, E. Pitchon, A. Bertsch, P. Renaud and A. Mermoud, "Wireless Contact Lens Sensor for Intraocular Pressure Monitoring: Assessment on Enucleated Pig Eyes," Acta Ophthalmologica, Vol. 87, No. 4, 2009, pp. 433-437.
24. R. Puers, "Capacitive Sensors: When and How to Use Them," Sensors and Actuators A, Vol. 37-38, 1993, pp. 93-105.
25. Y. Backlund, L. Rosengren, B. Hok and B. Svedbergh, "Passive Silicon Transensor Intended for Biomedical, Remote Pressure Monitoring," Sensor and Actuators A, Vol. 21-23, 1990, pp. 58-61.
26. K. Van Schuylenbergh and R. Pures, "Passive Telemetry by Harmonics Detection," Proceedings of 18th Annual International Conference, Vol. 1, Amsterdam, The Nederlands, pp. 299-300.
27. O. Akar, T. Akin and K. Najafi, "A Wireless Batch Sealed Absolute Capacitive Pressure Sensor," Sensors and actuators A, Vol. 95, 2001, pp. 29-38.
28. K. Stangel, S. Kolnsberg, D. Hammerschmidt, H. K. Trieu and W. Mokwa, "A Programmable Intraocular CMOS Pressure Sensor System Implant," IEEE Solid State Journal, Vol. 36, No. 7, 2001, pp. 1094-1100.

29. S. Lizon-Martinez, R. Gianetti, J. L. Rodriguez-Marrero and B. Tellini, "Design of a System for Continuous Intraocular Pressure Monitoring," IEEE Transactions on Instrumentation and Measurement, Vol. 54, No. 4, 2005, pp. 1534-1540.
30. E. Y. Chow, S. Chakraborty, W. J. Chapbell and P. P. Irazoqui, "Mixed-Signal Integrated Circuits for SelfContained Sub-Cubic Millimetre Biomedical Implants," International Solid State Circuits Conference, San Francisco, 2010.
31. C. Chouard and P. Pialoux, "Biocompatibility of Cochlear Implants," Bulletin National Academy of Medicine, Vol. 179, 1995, pp. 549-555.
32. D. R. Merrill, M. Bikson and J. Jefferys, "Electrical Simulation of Excitable Tissue: Design of Efficacious and Safe Protocols," Journal of Neuroscience Methods, Vol. 141, 2005, pp. 171-198.

Chapter 2

Design of Dual-Mode Substrate Integrated Waveguide Band-Pass Filters

Ahmed Rhbanou[1], Mohamed Sabbane[1], Seddik Bri[2]

[1]Department of Mathematics, FSM, Moulay Ismail University, Meknes, Morocco
[2]Material and Instrumentations Group (MIN), Electrical Engineering Department, ESTM, Moulay Ismail University, Meknes, Morocco

ABSTRACT

Three dual-mode band-pass filters are presented in the present paper. The first filter is realized by dual-mode substrate integrated waveguide (SIW) cavity; the second is based on the integration of SIW cavity with electromagnetic band gap (EBG); and the third is based on the integration of SIW cavity with complementary split ring resonator (CSRR). The dual-mode SIW cavity is designed to have a fractional bandwidth of 4.95% at the midband frequency of 9.08 GHz; the proposed EBG- SIW resonator operates at 9.12 GHz with a bandwidth of 4.38% and the CSRR-SIW resonator operates at 8.66 GHz with a bandwidth of 2.54%. The proposed filters have the high Q-factors and generate a transmission zero in upper stopband, and these by the use of Rogers RT/duriod 5880 (tm).

Keywords: Substrate Integrated Waveguide, Electromagnetic Band Gap, Complementary Split Ring Resonators, SIW Cavity, Transition

1. INTRODUCTION

Rectangular waveguide filters are widely used in RF-Microwave industry, due to its characteristic properties of low losses, and high Q-factor. However, their integrations with planar structures in electronic systems are very difficult and their fabrications are expensive.

To resolve these problems, a new technology is implemented, called the substrate integrated waveguide (SIW). The SIW responds to these constraints in the design of microwave components by taking the advantages of low radiation loss, high power handling and high Q-factor.

The SIW is formed by two solid conductor planes, separated by a dielectric substrate, with conductor sidewalls emulated by rows of metalized through-plated via [1] - [10] .

On the other hand, the metamaterials (electromagnetic band gaps (EBGs) and complementary split-ring resonators (CSRRs)) are used for manipulating electromagnetic waves and the unusual properties. The application of metamaterials has allowed a great improvement the performances and the size reduction in planar microwave applications, filter, and antenna [11] - [20] .

In this paper, three dual-mode band-pass filters are proposed. These filters have been simulated in a commercial software package HFSS™, and thus a comparison was made between the proposed filters and several previous filters reported in [21] -[24] .

2. DESIGN OF SIW CAVITY AND PROPOSED TRANSITIONS

The SIW cavity enables propagation of the mode (TE_{m0p}), its parameters necessary are the length (L_{SIW}), the width (W_{SIW}), the diameter D of the metallic via hole and the spacing P between the holes, which are expressed by the resonant frequency of rectangular cavity ($f_{c_{m0_p}}$), because the electrical behavior of SIW cavity is very close to rectangular cavity filled with the same dielectric (ε_r) of length (L_{eff}) and width (W_{eff}) as shown in Figure 1. The expression for the resonant frequency of SIW cavity:

$$f_{c_{m0p}} = \frac{c}{2\sqrt{\varepsilon_r}}\sqrt{\left(\frac{m}{W_{eff}}\right)^2 + \left(\frac{p}{L_{eff}}\right)^2}. \quad (1)$$

The size of SIW cavity is designed from the empirical Equations (1) and (2) These Equations are valid for $P < 4D$ and $P < \lambda_0 (\varepsilon_r/2)^{1/2}$ with λ_0 the space wavelength [1] -[9] .

$$W_{eff} = W_{SIW} - \frac{D^2}{0.95P} \tag{2}$$

$$L_{eff} = L_{SIW} - \frac{D^2}{0.95P} \tag{3}$$

The microstrip transition allows integration SIW to the planar structures (the planar transmission lines). Figure 2 shows the geometric structure of the proposed tapered transition. The parameters of the transition (L_T and W_T) and the microstrip line (W_M) are expressed from the relations in [10].

3. DESIGN OF META MATERIALS (EBG AND CSRR)

The electromagnetic band gap structures are presented as the complex periodic structures. The electromagnetic band gap materials are used for many applications and especially in the frequency filtering, because of their electromagnetic properties, to create band gaps in the electromagnetic spectrum. The application of EBG has allowed a great improvement the performances of numerous devices in telecommunication systems. Figure 3 shows the geometric structure of the proposed S-shape EBG.

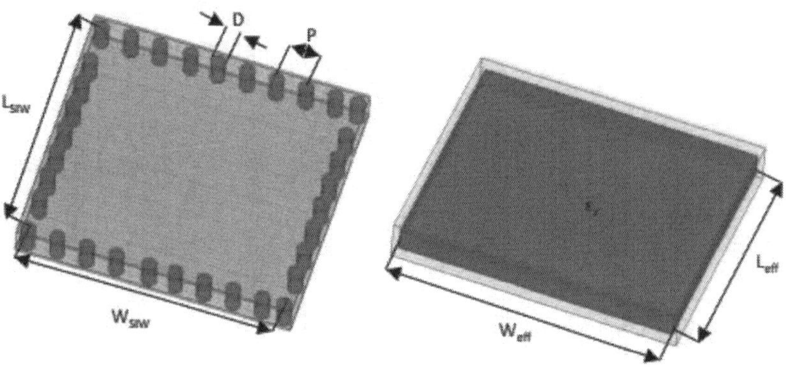

Figure 1. (a) SIW resonance cavity; (b) Conventional rectangular waveguide resonance cavity.

Where M is the length of EBG, T is the width of EBG and V is the strip-length of EBG. The EBG is founded on the Bragg condition [11] - [15] .

On the other, the CSRR is employed as LC resonator, thus the resonant frequency is obtained by the geometric parameters of CSRR. These specific properties are adaptable for many applications and especially the filters [16] - [20] . Figure 4 shows the geometric structure of the proposed CSRR cell. With W is the ring length of CSRR, G is the gap length, A is the side-length of CSRR and F is the width which separates two rings.

4. RESULTS

The dual-mode SIW cavity uses the substrate of Rogers RT/duriod 5880 (tm) (ε_r = 2.2, h = 0.508 mm and tanδ = 0.0009), with D = 0.6 mm and P = 1 mm. By considering two orthogonal modes TE_{102} and TE_{301}, the size of dual-mode SIW cavity is designed from the Equation (4).

$$W_{eff} = L_{eff}\sqrt{\frac{8}{3}} \tag{4}$$

The initial dimensions of dual-mode SIW cavity with tapered transitions have been optimized by software package HFSS™. The detailed dimensions are decided as: W_{SIW} = 37.4 mm, L_{SIW} = 22 mm, W_M = 1.568 mm, W_T = 8 mm and L_T = 14.22 mm. Figure 5 shows the geometric structure of dual-mode SIW cavity with tapered transitions.

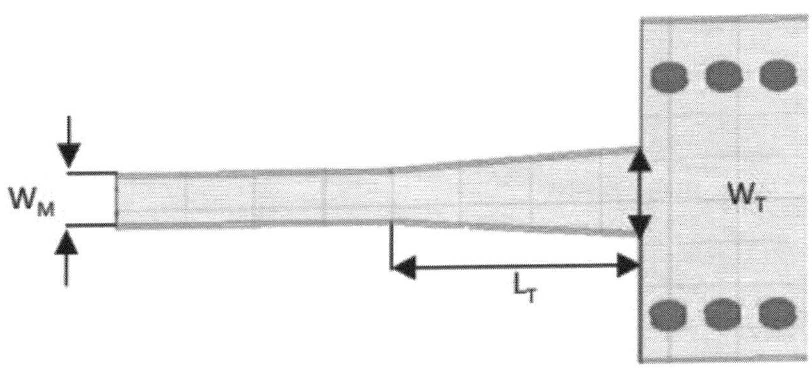

Figure 2. SIW with tapered transitions.

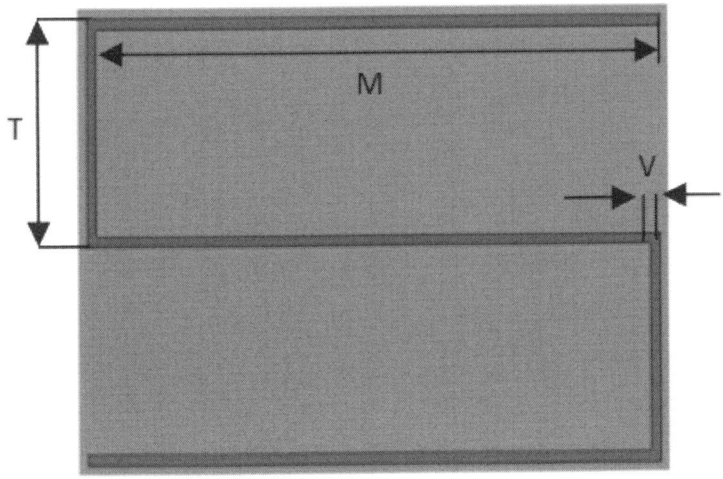

Figure 3. Geometry of S-shape EBG.

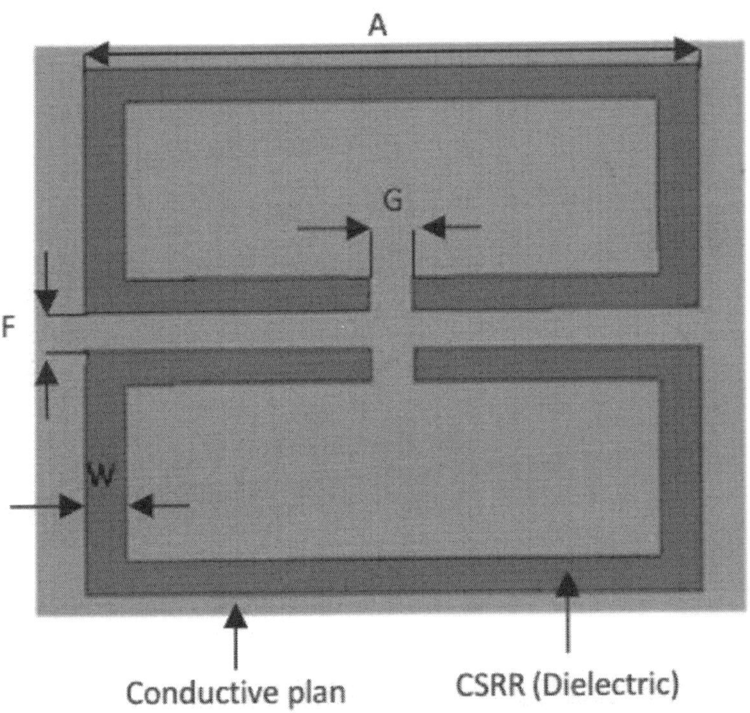

Figure 4. Geometry of CSRR.

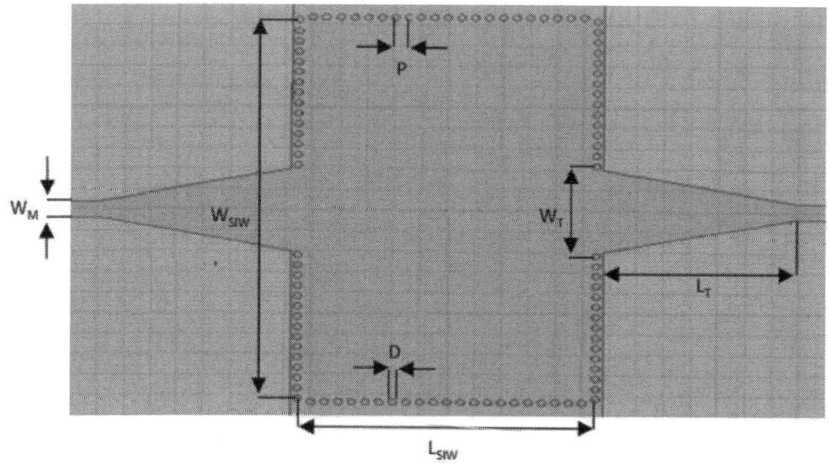

Figure 5. Configuration of dual-mode SIW cavity with tapered transitions.

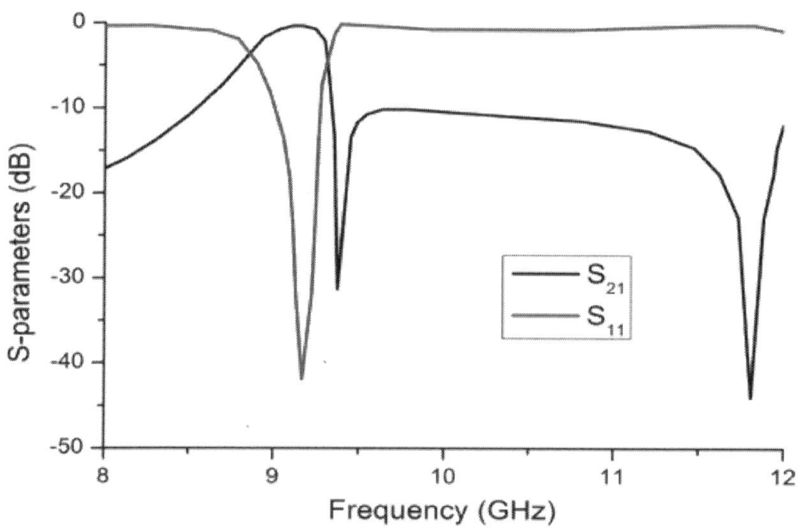

Figure 6. Simulated S-parameters of dual-mode SIW cavity with tapered transitions.

The simulation results for S-parameters of the dual-mode SIW cavity with tapered transitions are shown below in Figure 6.

Simulated results presented in Figure 6 show that the dual-mode SIW cavity with tapered transitions has the 3 dB bandwidth is approximately 0.45 GHz (from 8.855 to 9.305 GHz) centered at 9.08 GHz. The insertion loss is 0.43 dB and the return loss is better than 20 dB across the band of interest. Moreover, a transmission zeros at 9.38 GHz and the Q-factor is 414.

On the other hand, the integration of SIW cavity with EBG or CSRR allows the creation a dual-mode band-pass filter. The SIW is designed on substrate of Rogers RT/duriod 5880 (tm) (ε_r = 2.2, h = 0.508 mm and tanδ = 0.0009), with D = 0.6 mm and P = 1 mm. Figure 7 shows the proposed configuration of SIW-microstrip line with tapered transitions. The parameters of the proposed structure are optimized by software package HFSS™. The final desired dimensions are: W_{SIW} = 15 mm, W_M = 1.568 mm, L_T = 14.22 mm and W_T = 5.76 mm.

For analyzing the property of S-shaped EBG, one-cell EBG is etched on the top side of SIW. The dimensions of the one-cell EBG are: M = 2 mm, T = 5 mm, V = 0.9 mm. Figure 8 shows the configuration of one-cell SIW-EBG. The simulation results for S_{21} parameter of standard SIW and one-cell SIW-EBG are shown below in Figure 9.

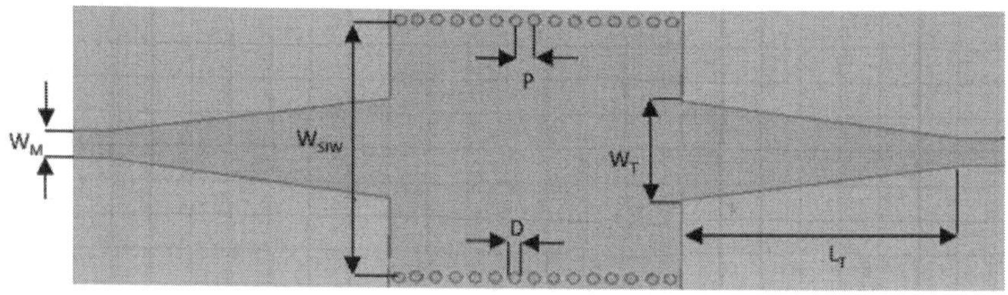

Figure 7. Configuration of SIW with tapered transitions.

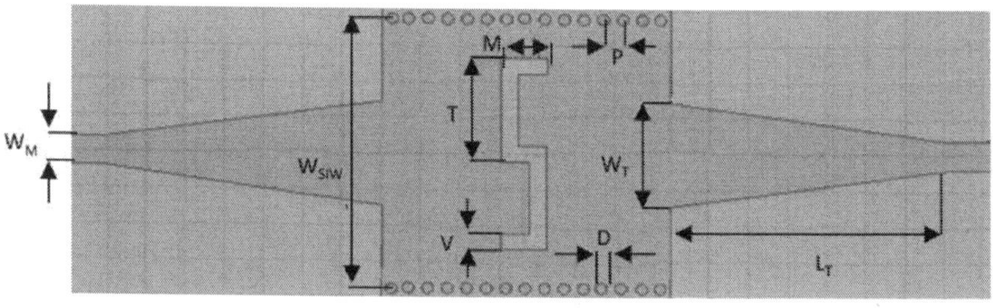

Figure 8. Configuration of one-cell SIW-EBG.

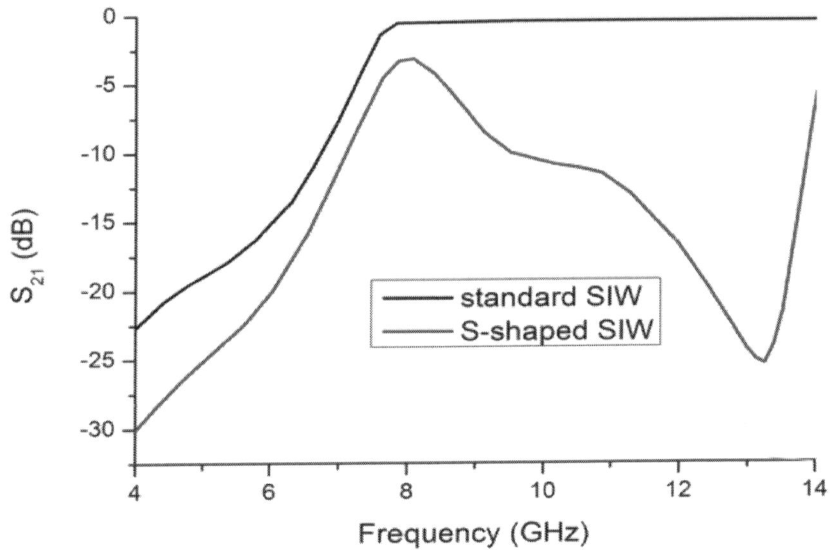

Figure 9. Simulated transmission coefficients S_{21} of standard SIW and one-cell SIW-EBG.

As shown in Figure 9, the structure of one-cell SIW-EBG has allowed producing a stopband at about 13.19 GHz.

On the other hand, Figure 10 shows the geometric structure of the proposed SIW resonator. The TE_{101}- mode-based SIW cavity is presented by a square cavity with $W_{SIW} = L_{SIW} =$ 15 mm and the tapered transitions with the same dimensions W_T = 5.76 mm, L_T = 14.22 mm and W_M = 1.568 mm.

The simulation results for S-parameters of the proposed SIW resonator are shown below in Figure 11. Simulated results presented in Figure 11 show that the SIW resonator has the 3 dB bandwidth is approximately 0.34 GHz at the midband frequency of 9.11 GHz, The insertion loss is 0.371 dB and the return loss is better than 20 dB across the band of interest. Moreover, the Q-factor is 641.

To make the characteristics of proposed SIW resonator clear, the width (W_{SIW}) is discussed in details. The simulation results for S_{21} parameter of the SIW resonator with different values of $W_{SIW}$are shown below in Figure 12. Table 1 shows the simulation results obtained.

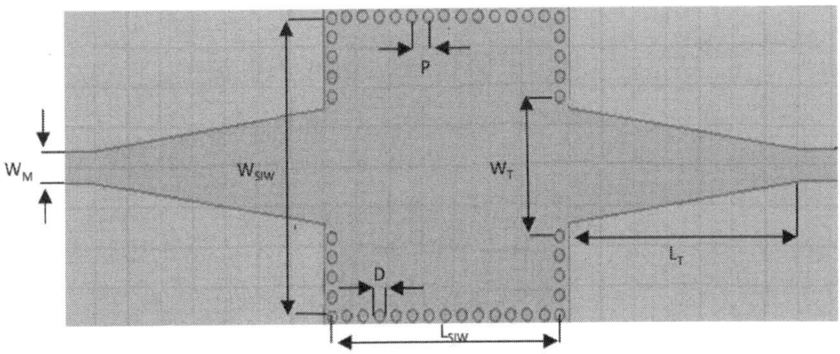

Figure 10. SIW resonator.

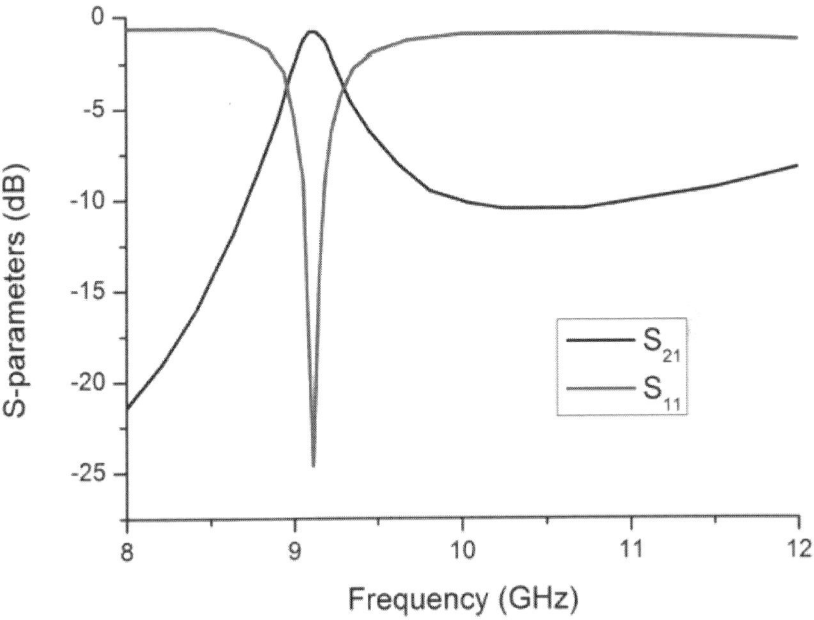

Figure 11. Simulated S-parameters of SIW resonator.

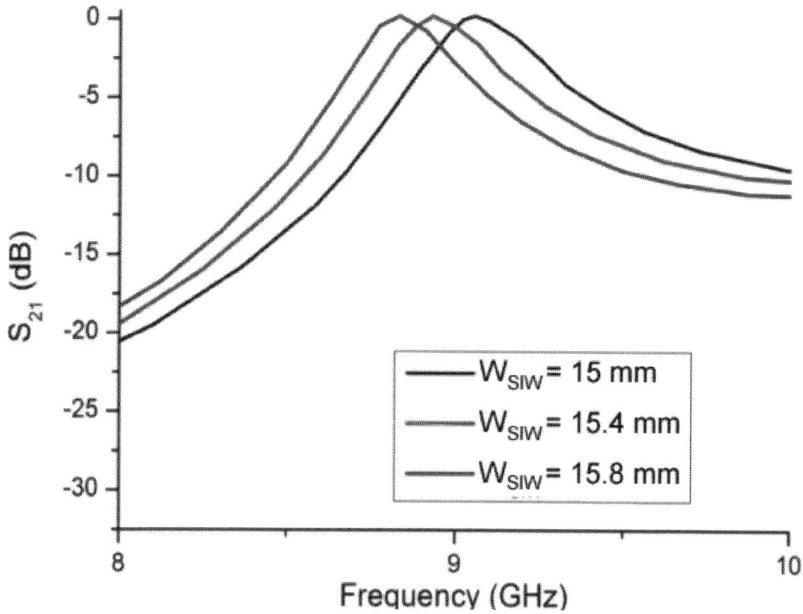

Figure 12. Simulated transmission coefficients S_{21} of SIW resonator with different values of W_{SIW}, where L_{SIW} = 15 mm, W_T = 5.76 mm, L_T = 14.22 mm, W_M = 1.568 mm.

Table 1. The simulated results of SIW resonator with different values of the width (W_{SIW}).

W_{SIW} (mm)	Passband frequency (GHz)	Center frequency (GHz)	Fractional bandwidth (%)	Insertion loss (dB)
15	8.94 to 9.27	9.11	3.73	0.371
15.4	8.84 to 9.17	8.98	3.67	0.4
15.8	8.74 to 9.05	8.87	3.49	0.42

As illustrated in Table 1, when the width (W_{SIW}) of SIW resonator increases the center frequency and the fractional bandwidth are decreasing, while the insertion loss becomes higher.

After studying the characteristics of SIW cavity and EBG, an EBG-SIW resonator is designed on the same substrate of Rogers RT/duriod 5880 (tm) (ε_r = 2.2, h = 0.508 mm and tanδ = 0.0009).Figure 13 shows the geometric structure of the proposed EBG-SIW resonator. Its physical parameters are provided in Table 2. The simulation results for S-parameters of the proposed EBG-SIW resonator are shown below in Figure 14.

Simulated results presented in Figure 14 show that the EBG-SIW resonator has the 3 dB bandwidth is approximately 0.4 GHz (from 8.92 to 9.32 GHz) centered at 9.12 GHz. The insertion loss is 1.18 dB and the return loss is better than 15 dB across the band of interest. Moreover, a transmission zeros at 12 GHz and the Q-factor is 179.

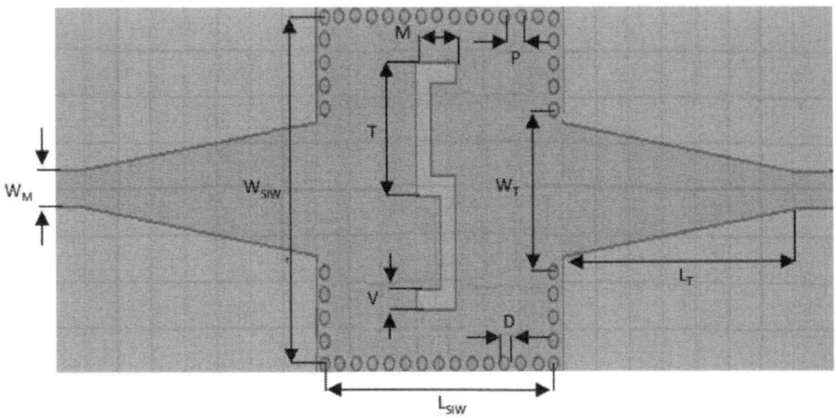

Figure 13. Geometry of EBG-SIW resonator.

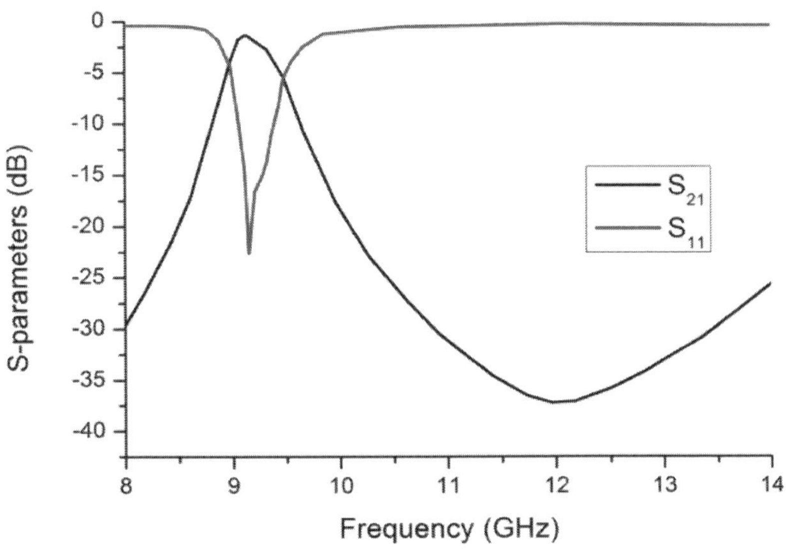

Figure 14. Simulated S-parameters of EBG-SIW resonator.

Table 2. Physical parameters of proposed EBG-SIW resonator.

Symbol	Value (mm)	Symbol	Value (mm)
D	0.6	L_T	14.22
P	1	W_M	1.568
W_{SIW}	15	M	2.4
L_{SIW}	15	T	5.4
W_T	5.76	V	0.9

On the other hand, the characteristics of CSRR are analyzed, by using a simple model as shown inFigure 15, this model is formed by etching a CSRR on the top side of SIW.

In order to make the characteristics of proposed CSRR clear, its side-length (A) is discussed in details, with the same conditions as ε_r = 2.2, h = 0.508 mm, W_T = 5.76 mm, W_M = 1.568 mm, L_T= 14.22 mm, W_{SIW} = 15 mm, D = 0.6 mm, P = 1 mm. The influence of the side-length (A) is simulated and shown in Figure 16. The simulation results are shown in Table 3.

Simulated results presented in Table 3 show that the resonant frequency decreases thus the attenuation increases when the side-length of CSRR becomes increasingly larger.

After studying the characteristics of CSRR, a CSRR-SIW resonator is designed on the same substrate of Rogers RT/duriod 5880 (tm) (ε_r = 2.2, h = 0.508 mm and tanδ = 0.0009). Figure 17 shows the geometric structure of CSRR-SIW resonator. Its physical parameters are provided in Table 4. The simulation results for S-parameters of the proposed CSRR-SIW resonator are shown below in Figure 18.

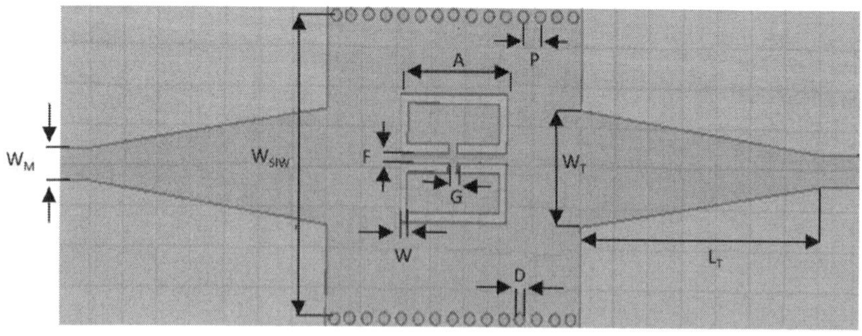

Figure 15. Configuration of one-cell SIW-CSRR.

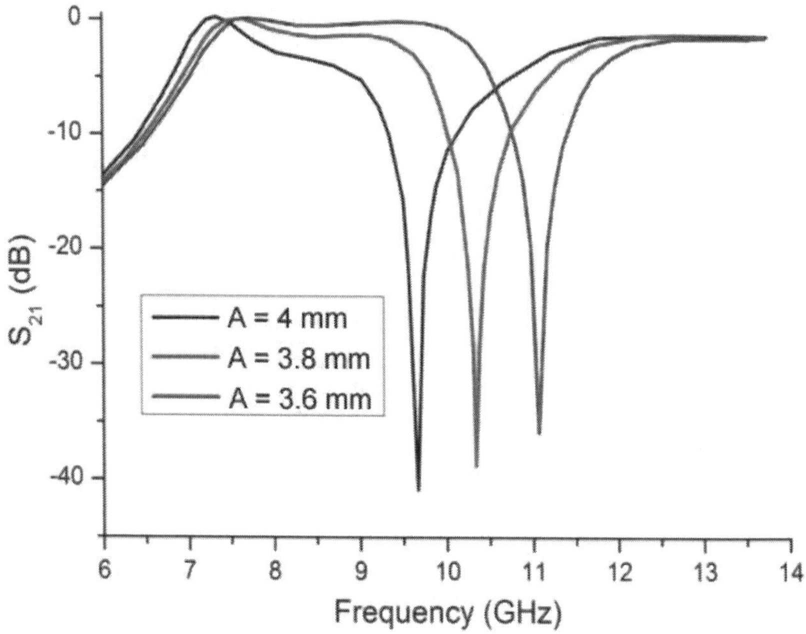

Figure 16. Simulated transmission coefficients S_{21} of one-cell SIW-CSRR with different values of A, where G = 0.3 mm, F = 0.3 mm, W = 0.3 mm.

Table 3. The simulated results of one-cell SIW-CSRR with different values of the side-length (A).

A (mm)	Centre stop band frequency (GHz)	Attenuation (dB)
3.6	11.06	36.66
3.8	10.32	38.84
4	9.65	40.88

Simulated results presented in Figure 18 show that the proposed CSRR-SIW resonator has the 3 dB bandwidth is approximately 0.22 GHz (from 8.55 to 8.77 GHz) centered at 8.66 GHz. The insertion loss is 0.55 dB and the return loss is better than 20 dB across the band of interest. Moreover, a transmission zeros at 11.73 GHz and the Q-factor is 640.

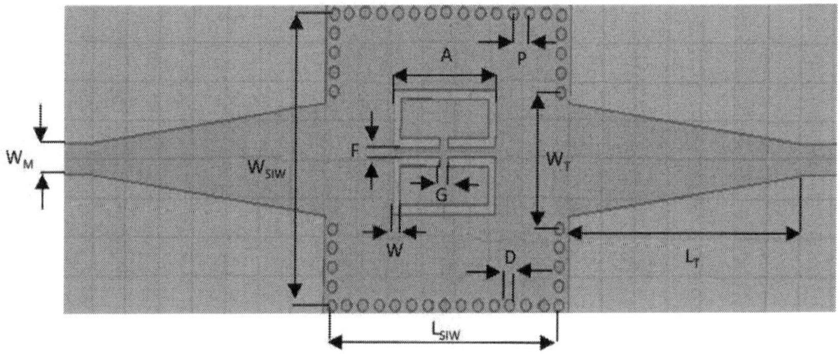

Figure 17. Geometry of CSRR-SIW resonator.

Table 4. Dimensions of proposed CSRR-SIW resonator.

Symbol	Value (mm)
D	0.6
P	1
W_{SIW}	15
L_{SIW}	15
W_T	5.76
L_T	14.22
W_M	1.568
A	3.5
F	0.3
W	0.3
G	0.3

In order to verify the characteristics of proposed filters, some comparisons between the proposed filters and several previous filters reported in the references are summarized in Table 5. According to the comparisons, the proposed filters have the advantage of high Q-factor compared to the filters in [21] -[24] .

5. CONCLUSIONS

In this paper, three dual-mode band-pass filters are proposed. The dual-mode SIW cavity filter has a center frequency of 9.08 GHz with a bandwidth of 4.95%. The insertion loss is 0.43 dB and the return loss is better than 20 dB across the band of interest. In addition, a transmission zero at 9.38 GHz and Q-factor is 414. The EBG-SIW resonator has a center frequency of 9.12 GHz with a bandwidth of 4.38%. The insertion loss is 1.18 dB and the return loss is better than 15 dB across the band of interest. In addition, a transmission zero at 12 GHz and Q-factor is 179. The CSRR-SIW resonator has a center frequency of 8.66 GHz with a bandwidth of 2.54%. The insertion loss is 0.55 dB and the return loss is better than 20 dB across the band of interest. In addition, a transmission zero at 11.73 GHz and Q-factor is 640. The simulation processes of the structures are done by using HFSS software. The design methods are discussed and presented.

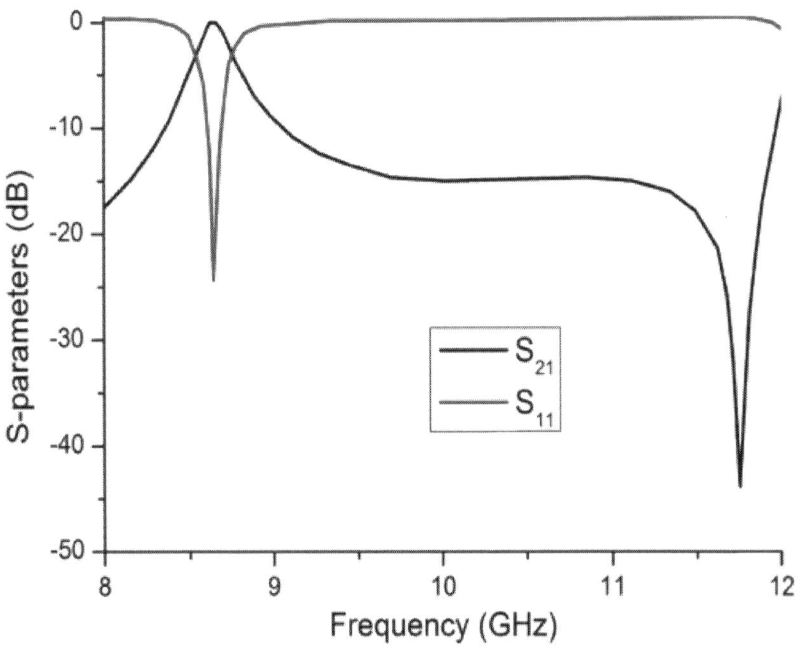

Figure 18. Simulated S-parameters of CSRR-SIW resonator.

Table 5. Performance comparison among published dual-mode band pass filters and the proposed filters.

Dual-mode band pass filters	Centre frequency (GHz)	Fractional bandwidth (%)	Insertion loss (dB)	Q-factor
[21]	2.41	4.50	1.38	151
[22]	2.40	12.50	0.39	182
[23]	14.42	2.64	2.60	146
[24]	5.00	4.00	0.70	322
Dual-mode SIW cavity	9.08	4.95	0.43	414
EBG-SIW resonator	9.12	4.38	1.18	179
CSRR-SIW resonator	8.66	2.54	0.55	640

The proposed filters have a small size, high Q-factor and low loss, and can be directly integrated with other circuits without any additional mechanical assembling tuning. Additionally, these filters are easily scalable over microwave and millimeter frequency ranges.

REFERENCES

1. Cassivi, Y., Perregrini, L., Arcioni, P., Bressan, M., Wu, K. and Conciauro, G. (2002) Dispersion Characteristics of Substrate Integrated Rectangular Waveguide. IEEE Microwave and Wireless Components Letters, 12, 333-335.
2. Deslandes, D. and Wu, K. (2006) Accurate Modeling, Wave Mechanism, and Design Consideration of a Substrate Integrated Waveguide. IEEE Transactions on Microwave Theory and Techniques, 54, 2516-2526.
3. Djerafi, T. and Wu, K. (2007) Super-Compact Substrate Integrated Waveguide Cruciform Directional Coupler. IEEE Microwave and Wireless Components Letters, 17, 757-759.
4. Su, P., Tang, Z.-X. and Zhang, B. (2012) Push-Push Dielectric Resonator Oscillator Using Substrate Integrated Waveguide Power Combiner. Progress in Electromagnetics Research Letters, 30, 105-113.
5. Huang, Y., Shao, Z. and Liu, L. (2013) A Substrate Integrated Waveguide Bandpass Filter Using Novel Defected Ground Structure Shape. Progress in Electromagnetics Research, 135, 201-213.
6. Zhang, X.-C., Yu, Z.-Y. and Xu, J. (2007) Novel Band-Pass Substrate Integrated Waveguide (SIW) Filter Based on Complementary Split Ring Resonators (CSRRS). Progress in Electromagnetics Research, 72, 39-46.

7. Xu, F. and Wu, K. (2005) Guided-Wave and Leakage Characteristics of Substrate Integrated Waveguide. IEEE Transactions on Microwave Theory and Techniques, 53, 66-73.
8. Cassivi, Y., Perregrini, L., Wu, K. and Conciauro, G. (2002) Low-Cost and High-Q Millimeter-Wave Resonator Using Substrate Integrated Waveguide Technique. The 32nd European Microwave Conference, Milan, 23-26 September 2002, 1-4.
9. Rhbanou, A., Bri, S. and Sabbane, M. (2015) Design of X-Band Substrate Integrated Waveguide Bandpass Filter with Dual High Rejection. Microwave and Optical Technology Letters, 57, 1744-1752.
10. Deslandes, D. (2010) Design Equations for Tapered Microstrip-to-Substrate Integrated Waveguide Transitions. 2010 IEEE MTT-S International Microwave Symposium Digest (MTT), Anaheim, 23-28 May 2010, 704-707.
11. Kyriazidou, C.A., Contopanagos, H.F. and Alexopoulos, N.G. (2001) Monolithic Waveguide Filters Using Printed Photonic-Bandgap Materials. IEEE Transactions on Microwave Theory and Techniques, 49, 297-307.
12. Falcone, F., Lopetegi, T. and Sorolla, M. (1999) 1-D and 2-D Photonic Bandgap Microstrip Structures. Microwave and Optical Technology Letters, 22, 411-412.
13. Garcia-Garcia, J., Bonache, J., Gil, I., Martin, F., Marqués, R., Falcone, F., Lopetegi, T., Laso, M.A.G. and Sorolla, M. (2005) Comparison of Electromagnetic Band Gap and Split-Ring Resonator Microstrip Lines as Stop Band Structures. Microwave and Optical Technology Letters, 44, 376-379.
14. Hao, Z.-C., Hong, W., Chen, J.-X., Chen, X.-P. and Wu, K. (2005) Compact Super-Wide Bandpass Substrate Integrated Waveguide (SIW) Filters. IEEE Transactions on Microwave Theory and Techniques, 53, 2968-2977.
15. Li, D., Tong, C.-M., Bao, J.-S., Peng, P. and Yu, D.-W. (2013) A Novel Bandpass Filter of Substrate Integrated Waveguide (SIW) Based on S-Shaped EBG. Progress in Electromagnetics Research Letters, 36, 191-200.
16. Bonache, J., Gil, I., Garcia-Garcia, J. and Martin, F. (2006) Novel Microstrip Bandpass Filters Based on Complementary Split-Ring Resonators. IEEE Transactions on Microwave Theory and Techniques, 54, 265-271.

17. Dong, Y.D., Yang, T. and Itoh, T. (2009) Substrate Integrated Waveguide Loaded by Complementary Split-Ring Resonators and Its Applications to Miniaturized Waveguide Filters. IEEE Transactions on Microwave Theory and Techniques, 57, 2211-2223.
18. O'Brien, S. and Pendry, J.B. (2002) Magnetic Activity at Infrared Frequencies in Structured Metallic Photonic Crystals. Journal of Physics: Condensed Matter, 14, 6383-6394.
19. Baena, J.D., Bonache, J., Martin, F., Sillero, R.M., Falcone, F., Lopetegi, T., Laso, M.A.G., Garcia-Garcia, J., Gil, I., Portillo, M.F. and Sorolla, M. (2005) Equivalent-Circuit Models for Split-Ring Resonators and Complementary Split-Ring Resonators Coupled to Planar Transmission Lines. IEEE Transactions on Microwave Theory and Techniques, 53, 1451-1461.
20. Bilotti, F., Toscano, A. and Vegni, L. (2007) Design of Spiral and Multiple Split Ring Resonators for the Realization of Miniaturized Metamaterial Samples. IEEE Transactions on Antennas and Propagation, 55, 2258-2267.
21. Lin, Y.-F., Chen, C.-H., Chen, K.-Y., Chen, H.-M. and Wong, K.-L. (2007) A Miniature Dual-Mode Bandpass Filter Using Al2O3 Substrate. IEEE Microwave and Wireless Components Letters, 17, 580-582.
22. Huang, X.D. and Cheng, C.H. (2006) A Novel Coplanar-Waveguide Bandpass Filter Using a Dual-Mode Square-Ring Resonator. IEEE Microwave and Wireless Components Letters, 16, 13-15.
23. Li, R.Q., Tang, X.H. and Xiao, F. (2010) Substrate Integrated Waveguide Dual-Mode Filter Using Slot Lines Perturbation. Electronics Letters, 46, 845-846.
24. Wu, L.-S., Zhou, X.-L., Wei, Q.-F. and Yin, W.-Y. (2009) An Extended Doublet Substrate Integrated Waveguide (SIW) Bandpass Filter with a Complementary Split Ring Resonator (CSRR). IEEE Microwave and Wireless Components Letters, 19, 777-779.

Chapter 3

Fully Integrated Low-Noise Readout Circuit with Automatic Offset Cancellation Loop for Capacitive Microsensors

Haryong Song[1], Yunjong Park[1], Hyungseup Kim[1], Dong-il Dan Cho[2] and Hyoungho Ko[1,*]

[1]Department of Electronics, Chungnam National University, Daejeon 305-764, Korea
[2]ASRI/ISRC, Department of Electrical Engineering and Computer Science, Seoul National University, Seoul 151-744, Korea

ABSTRACT

Capacitive sensing schemes are widely used for various microsensors; however, such microsensors suffer from severe parasitic capacitance problems. This paper presents a fully integrated low-noise readout circuit with automatic offset cancellation loop (AOCL) for capacitive microsensors. The output offsets of the capacitive sensing chain due to the parasitic capacitances and process variations are automatically removed using AOCL. The AOCL generates electrically equivalent offset capacitance and enables charge-domain fine calibration using a 10-bit R-2R digital-to-analog converter, charge-transfer switches, and a charge-storing capacitor. The AOCL cancels the unwanted offset by binary-search algorithm based on 10-bit successive approximation register (SAR) logic. The chip is implemented using 0.18 μm complementary metal-oxide-semiconductor (CMOS) process with an active area of 1.76 mm2. The power consumption is 220 μW with 3.3 V supply. The input parasitic capacitances within the range of −250 fF to 250 fF can be cancelled out automatically, and the required calibration time is lower than 10 ms.

Keywords: capacitive sensing circuit; automatic offset cancellation loop (AOCL); correlated double sampling (CDS); capacitive microsensor

1. INTRODUCTION

Capacitive microsensors based on microelectromechanical system (MEMS) technologies are used in various application areas, including accelerometers, gyroscopes, pressure sensors, touch-screen sensors, proximity sensors, and so on because of their small form factor, low-power characteristics, good temperature dependency, and low cost [1,2,3,4,5]. The capacitive sensing method, however, suffers from severe parasitic capacitance mismatches. The parasitic capacitance mismatches are often much higher (on the order of several hundreds of femtofarads to several picofarads) than sensing capacitance charges (on the order of several tens of femtofarads). The parasitic capacitance mismatches and random process variations result in large output offset variations, and in worst case, output saturation to VDD or GND. Thus, additional offset calibration steps are required using external calibration equipment. Moreover, to calibrate the output offset of the physical sensors such as accelerometer, special equipment with physical stimulus capability, such as a vibration exciter, is required. These physical calibration steps increase the production cost of the sensor system. For example, in case of BMC050 six-axis combo motion sensor of Bosch Sensortec, test and packaging cost is 35% of the production cost [6]. The on-chip automatic offset calibration circuit using binary-weighted capacitive digital-to-analog converter was reported [7]. In the previous research [7], the calibration resolution of the offset calibrator is limited by the minimum design rules, thus, accurate automatic offset calibration with sub-femtofarads resolution was not implemented.

This paper presents a fully integrated low-noise capacitive sensing circuit with automatic offset calibration loop (AOCL). The output offsets of the capacitive sensing chain due to the parasitic capacitances and process variations are automatically removed using the automatic offset cancellation loop (AOCL). The AOCL generates an electrically equivalent offset capacitance, and it enables charge-domain fine calibration using a 10-bit R-2R digital-to-analog converter, charge-transfer switches and a charge-storing capacitor. The presented circuit enables the automatic offset calibration for capacitive sensors without using external test equipment.

2. CIRCUIT DESCRIPTION

2.1. Top Level Architecture

Figure 1 shows a block diagram of the proposed capacitive sensing circuit. The capacitive sensing chain adopts the correlated double sampling (CDS) technique to reduce the low frequency noise, including the 1/f noise [8,9,10]. The capacitive sensing chain is composed of three amplification stages. In the first stage, a capacitive sensing amplifier (CSA) converts the input capacitance change from MEMS sensor to output voltage. In the second stage, a programmable gain amplifier (PGA) amplifies the signal by 30 dB. The single-to-differential amplifier (SDA) of third stage converts the single-ended output of the PGA into a differential signal. The differential signals are converted to digital signals using a 12-bit successive approximation register (SAR) analog-to-digital converter (ADC) [11] after low pass filters and buffers. The 4th order Bessel low pass filter is implemented using continuous-time Sallen-Key active filter topology. The cut-off frequency of the low pass filter is 400 Hz. The AOCL adjusts the output offset to the desired value using binary search algorithm [12], and is implemented using a comparator, SAR logic, R-2R digital-to-analog converter (DAC), charge transfer switches, and a charge-storing capacitor.

2.2. Capacitive Sensing Amplifier (CSA)

Figure 2a,b show a correlated double sampled (CDS) CSA with offset calibration using conventional capacitor arrays and charge-domain fine calibration, respectively. The CDS operation removes unwanted low frequency noise components including flicker noise and DC offset. In conventional CSA, as shown in Figure 2a, binary-weighted capacitor arrays are used to cancel the input parasitic capacitance [13]. The resolution of conventional offset calibration scheme using capacitor arrays is limited by the minimum design rules of capacitors. For example, the minimum allowable capacitance is typically 16 fF (4 μm × 4 μm × 1 fF/μm^2) for a general metal-insulator-metal (MIM) capacitor. Assuming a sensing capacitance of several femtofarads, to achieve the high offset accuracy, offset calibrations with sub-fF steps are required. In this design, to implement a calibration capacitance smaller than the physical design rule, charge-domain fine calibration scheme is implemented, as shown in Figure 2b [13]. The comparisons between the conventional capacitor array calibration scheme and the charge-domain calibration scheme are summarized in Table 1. The output voltage using charge-domain calibration (V_o) and charge-domain equivalent capacitance (C_{eq}) can be expressed as Equation (1). In this design, the charge storing capacitor, C_{off}, is 250 fF, and the 10-bit R-2R DAC generates voltage from GND to VDD in 1024 steps. Thus, the charge-domain calibration circuit can generate an electrically equivalent offset capacitance in the range of ±250 fF with 0.488 fF step. The digital inputs of R-2R DAC are automatically determined by binary search algorithm of AOCL.

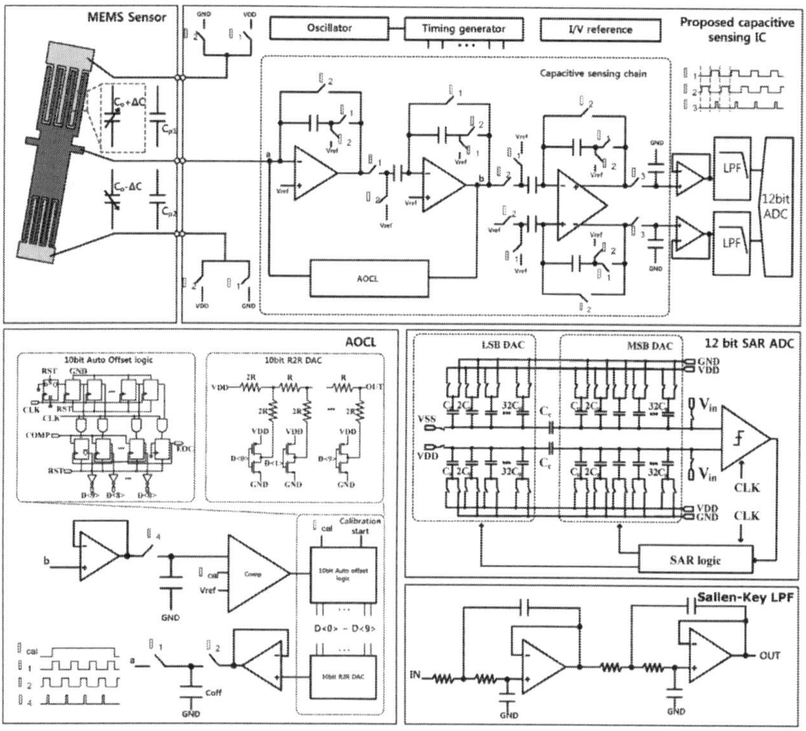

Figure 1. The block diagram of the proposed capacitive sensing circuit.

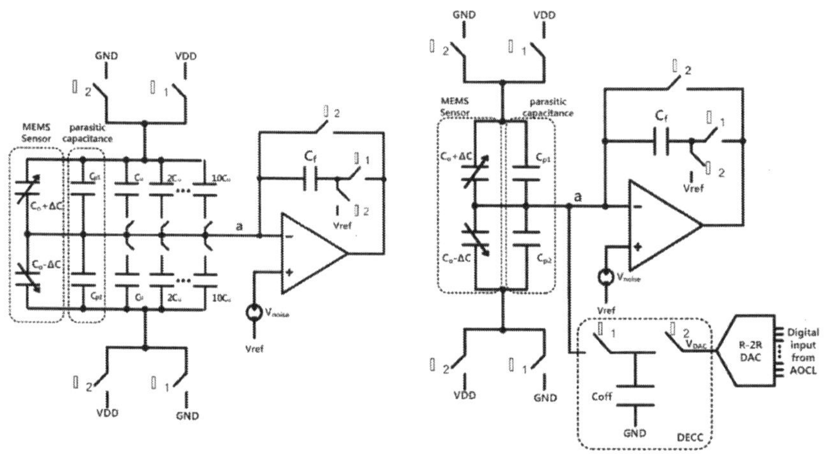

Figure 2. Capacitive sensing amplifier, (**a**) Conventional capacitor arrays calibration; (**b**) Charge-domain calibration.

$$V_o = -\frac{V_{DD}(2\Delta C + (C_{p1} - C_{p2}) + C_{eq})}{C_f} + V_{ref}$$

$$C_{eq} = \frac{(V_{ref} - V_{DAC})}{V_{DD}} \times C_{off} \tag{1}$$

Table 1. Comparisons between capacitor arrays calibration and charge-domain calibration.

Capacitor Implementation	Capacitor Arrays Calibration	Charge-Domain Calibration
	Physical Capacitor (MIM or PIP)	Electrically Equivalent Capacitor
Minimum capacitor	Limited by physical design rules (in this design, 16 fF = 4 μm × 4 μm × 1 fF/μm²)	LSB voltage × C_{off}/V_{DD} (in this design, 0.488 fF = 1/1024 × 3.3/3.3 × 500 fF)
Size	Large (binary-weighted capacitor array)	Small (R-2R DAC, switches, and a charge-storing capacitor)
DC current	0	DC current consumption in R-2R DAC (1.2 μA in this design)

2.3. Programmable Gain Amplifier (PGA) and Single-to-Differential Amplifier (SDA)

Figure 3 shows the PGA and SDA circuit. The PGA and SDA employ correlated double sampling technique to reduce the low-noise components. The Φ1 and

Φ2 are non-overlapping clocks. In PGA, an input coupling capacitor (C_{in1}) stores the input voltage signal and noise in Φ1 phase, and the stored charge is dumped in a feedback capacitor in Φ2 phase. The gain of PGA can be adjusted from 0 dB to 30 dB by programming C_{f1}, as expressed in Equation (2). SDA works with the opposite phase of PGA, and the differential gain of SDA can be adjusted from 0 dB to 36 dB by programming C_{f2}, as expressed in Equation (3). The output signal from the SDA is sampled on Φ3 phase.

$$V_{PGA} = V_{ref} + V_{in1} \frac{C_{f1}}{C_{in1}} \tag{2}$$

$$V_{S2D} = V_{ref} + 2V_{in2} \frac{C_{f2}}{C_{in2}} \tag{3}$$

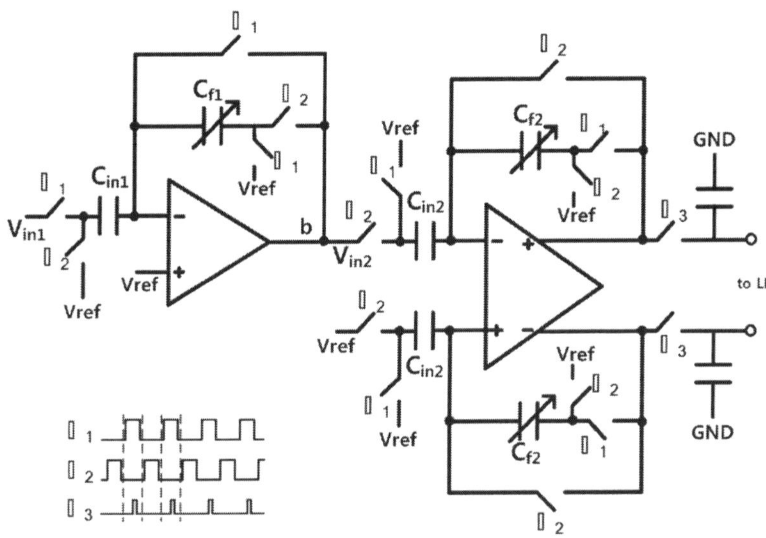

Figure 3. Programmable gain amplifier (PGA) and single to differential amplifier (SDA) circuit.

2.4. Automatic Offset Cancellation Loop (AOCL)

The proposed AOCL circuit is shown in Figure 4. The output voltage of PGA is sampled in Φ4 phase. In the 10-bit binary search successive approximation register (SAR) logic, the shift registers in the first row sequentially point towards the registers in the second row, which are to be updated to the comparator output. The registers (D-flip-flops) in the second row of the SAR logic become the digital control input of the R-2R digital to analog converter (DAC).

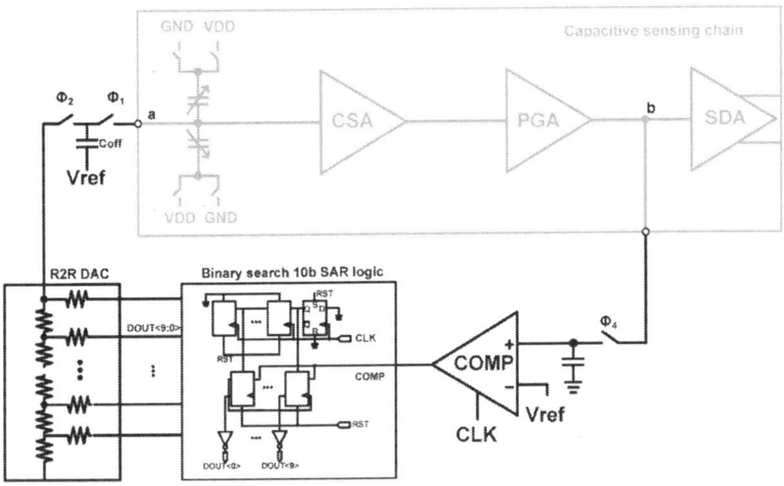

Figure 4. The proposed automatic offset cancellation loop (AOCL).

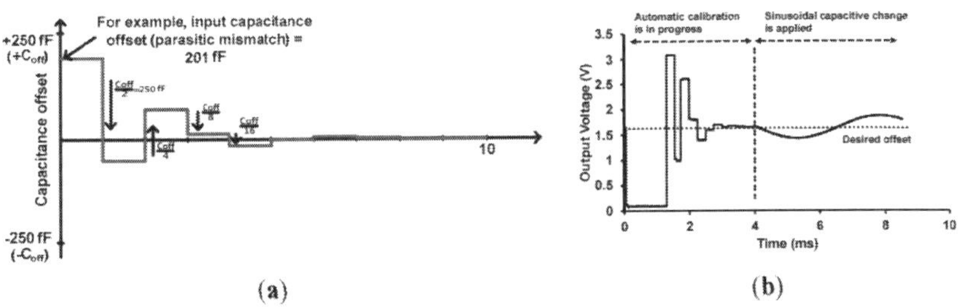

Figure 5. Operation example and simulation results of automatic offset cancellation loop (AOCL), (**a**) Operation example of AOCL; (**b**) Simulation result of AOCL.

An operation example of AOCL is illustrated in Figure 5a, and the steps involved in the operation are as follows. The sampled voltage is compared to Vref (desired voltage, VDD/2 = 3.3 V/2 = 1.65 V). Initially the MSB of DOUT<9:0> is “H”, and the initial value of DOUT<9:0> is “1,000,000,000”. With this initial value, the sampled PGA output and Vref are compared. If PGA output is higher than Vref, the MSB of DOUT<9:0> remains “H”, and DOUT<9:0> becomes “1,000,000,000”. If PGA output is lower than Vref, the MSB of DOUT<9:0> becomes “L”, and DOUT<9:0> becomes “0000000000”. DOUT<9:0> determines the digital input of R-2R DAC. The R-2R DAC, charge transfer switches, and charge storing capacitor generate an electrically equivalent offset capacitance. Next, the second bit of DOUT<9:0> is set to “H”, and DOUT<9:0> becomes “C100,000,000”, where C means the output of comparator. The PGA output is compared to Vref again, and the second bit of DOUT<9:0> is updated with new comparator output. After ten comparison cycles, the PGA output tracks the desired output of Vref, as shown in Figure 5b. The AOCL can calibrate the offset due to the input capacitive mismatches of MEMS sensors from −250 fF to 250 fF with 1024 steps.

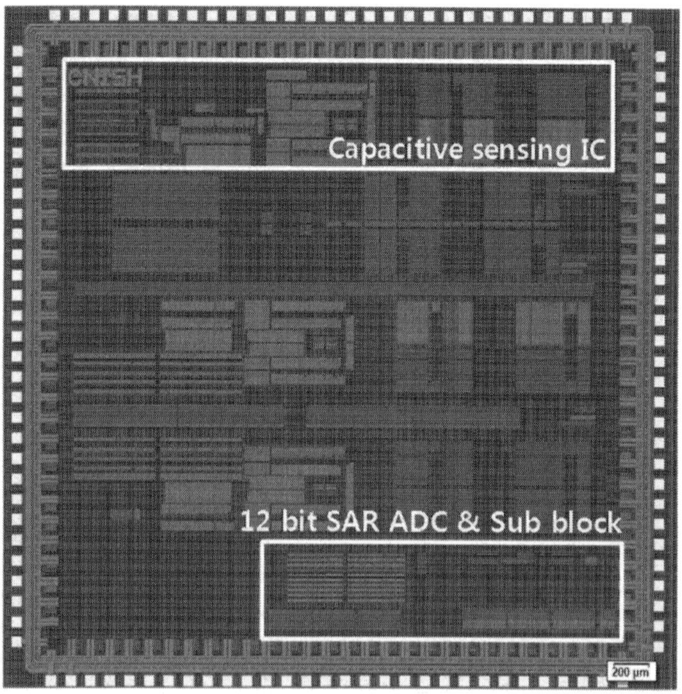

Figure 6. Chip micrograph.

3. MEASUREMENT RESULTS

A die-photograph of the fabricated capacitive sensing IC is shown in Figure 6. The IC is fabricated using a 0.18 μm single-polysilicon six-metal complementary metal-oxide-semiconductor (CMOS) process with an active area of 1.76 mm^2. Figure 7shows the measured analog output waveform with the AOCL operation. The green line and blue line show the differential output of the low pass filter. The yellow line is enable signal of AOCL, and the AOCL is activated at the falling edge of enable signal. After the operation of AOCL, the initial offset is removed automatically. The offset calibration time is 10 ms with 1 kHz calibration clock. Because the AOCL is operated before the low pass filter, the calibration clock for AOCL can be faster than cut-off frequency of the low pass filter. The frequency of calibration clock and cut-off frequency of the low pass filter are also programmable. The default cut-off frequency of the low pass filter is 400 Hz.

To evaluate the performance of the fabricated IC, the IC and MEMS *Z*-axis capacitive accelerometer are mounted on the vibration exciter, as shown in Figure 8. The detailed design and specification of the Z-axis accelerometer is described in [14].

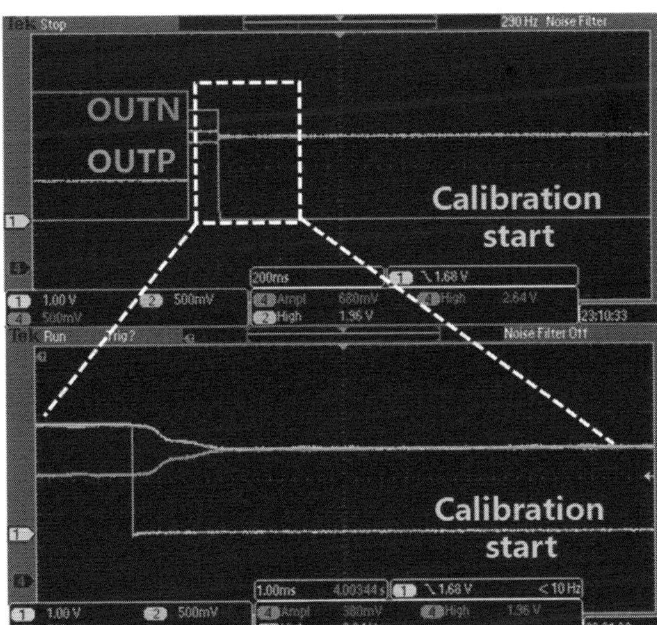

Figure 7. Measured analog output waveform with AOCL operation.

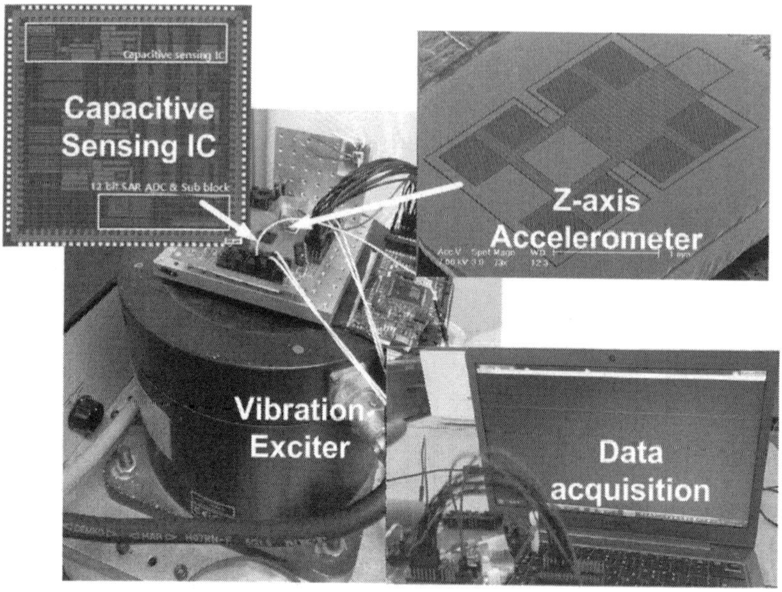

Figure 8. Measurement setup with capacitive accelerometer.

The measurement results are presented in Figure 9. Figure 9a shows the input referred noise spectrum. The input referred capacitance noise density, and integrated noise with 400 Hz bandwidth are 1.275 aF/√Hz and 25.5 aF_{RMS}, respectively. The input-output characteristics with MEMS *Z*-axis accelerometer are presented in Figure 9b. The scale factor, input range and, non-linearity are 0.1892 V/g, ±7.5 g, and 0.81% FSO, respectively.

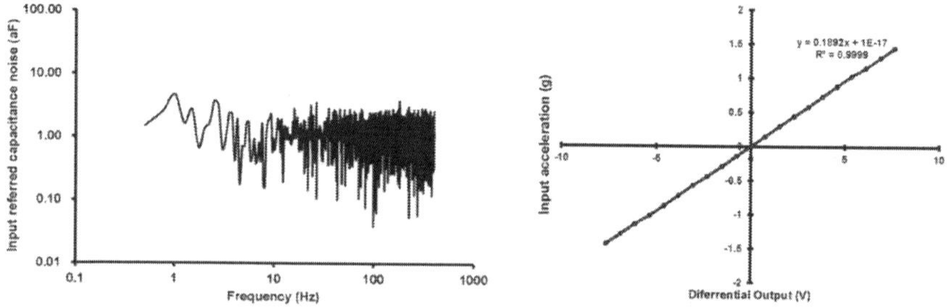

(a) (b)

Figure 9. Measurement results, (**a**) Input-referred noise; (**b**) Input-output characteristics.

4. CONCLUSIONS

A fully integrated low-noise capacitive readout circuit with automatic offset cancellation loop (AOCL) for capacitive microsensors has been presented. The output offsets of the capacitive sensing chain due to the parasitic capacitances and process variations were automatically removed using the AOCL. The AOCL cancelled the unwanted offset by binary-search algorithm based on 10-bit SAR logic and charge-domain calibration circuits. The chip was implemented using 0.18 μm 1P6M CMOS process with an active area of 1.76 mm^2. The power consumption was 220 μW with 3.3 V supply. The input parasitic capacitances within the range of −250 fF to 250 fF were cancelled out automatically, and the required calibration time was lower than 10 ms. The input referred capacitance noise density and integrated noise with 400 Hz bandwidth were 1.275 aF/√Hz and 25.5 aF_{RMS}, respectively. The scale factor, input range, and non-linearity were 0.1892 V/g, ±7.5 g, and 0.81% FSO, respectively. The presented circuit enabled the automatic offset calibration for capacitive sensors without using external test equipment, and enabled low-cost manufacturing of the capacitive sensors.

ACKNOWLEDGMENTS

This research was supported by a grant to Bio-Mimetic Robot Research Center Funded by Defense Acquisition Program Administration, and by Agency for Defense Development (UD130070ID). This work was also supported by IC Design Education Center (IDEC).

AUTHOR CONTRIBUTIONS

Haryong Song is first author, who has implemented the IC and written the draft of the manuscript. Yunjong Park and Hyungseup Kim have helped with the circuit design, simulation, layout, and evaluation. Dong-il Dan Cho has helped with the system design, performance evaluations, and testing of the MEMS accelerometer. Hyoungho Ko is the key inventor of the idea, who has designed the initial version of the IC, and administrated the overall project.

REFERENCES

1. Sun, H.; Fang, D.; Jia, K.; Maarouf, F.; Qu, H.; Xie, H. A low-power low-noise dual-chopper amplifier for capacitive CMOS-MEMS accelerometers. *IEEE Sens. J.* **2011**, *11*, 925–633.
2. Tez, S.; Akin, T. Fabrication of a Sandwich Type Three Axis Capacitive MEMS Accelerometer. In Proceedings of the IEEE Sensors, Baltimore, MD, USA, 3–6 November 2013; pp. 1–4.
3. Yu, H.Y.; Qin, M.; Huang, J.Q.; Huang, Q.A. A MEMS Capacitive Pressure Sensor Compatible with CMOS Process. In Proceedings of the IEEE Sensors, Taipei, Taiwan, 28–31 October 2012; pp. 1–4.
4. Tan, Z.; Shalmany, S.H.; Meijer, G.; Pertijs, M.A. An energy-efficient 15-bit capacitive-sensor interface based on period modulation. *IEEE J. Solid-State Circ.* **2012**, *47*, 1703–1711.
5. Dobrzynska, J.A.; Gijs, M.A.M. Polymer-based flexible capacitive sensor for three-axial force measurements. *J. Micromech. Microeng.* **2013**, *23*, 1–11.
6. Yole Developpement & System Plus Consulting, Inertial MEMS Manufacturing Trends 2014. Available online: http://www.systemplus.fr/reverse-costing-reports/inertial-mems-inertial-mems-manufacturing-trends-2014-volumes-1-2(accessed on 10 October 2015).
7. Shin, D.; Lee, H.; Kim, S. Delta–sigma interface circuit for capacitive sensors with an automatically calibrated zero point. *IEEE Trans. Circuits Syst. II Exp. Briefs.* **2011**, *58*, 90–94.
8. Kim, G.; Seok, C.; Kim, T.; Park, J.H.; Kim, H.; Ko, H. The Micro Pirani Gauge with Low Noise CDS-CTIA for In-Situ Vacuum Monitoring. *J. Semicond. Tech. Sci.* **2014**, *14*, 733–740.
9. Ha, H.; Suh, Y.; Lee, S.; Park, H.; Sim, J. A 0.5-V, 11.3-μW, 1-kS/s Resistive Sensor Interface Circuit with Correlated Double Sampling. In Proceedings of the IEEE Custom Integrated Circuits Conference (CICC), San Jose, CA, USA, 9–12 September 2012; pp. 1–4.
10. Perenzoni, M.; Massari, N.; Stoppa, D.; Pancheri, L.; Malfatti, M.; Gonzo, L. A 160 by 120-pixels range camera with in-pixel correlated double sampling and fixed-pattern noise correction. *IEEE J. Solid-State Circ.* **2011**, *46*, 1672–1681.
11. Wong, S.; Chio, U.; Zhu, Y.; Sin, S.; Seng-Pan, U.; Martins, R. A 2.3 mW 10-bit 170 MS/s two-step binary-search assisted time-interleaved SAR ADC. *IEEE J. Solid-State Circ.* **2013**, *48*, 1783–1794.

12. Zhand, D.; Bhide, A.; Alvandpour, A. A 53-nW 9.1-ENOB 1-kS/s SAR ADC in 0.13-m CMOS for medical implant devices. *IEEE J. Solid-State Circ.* **2012**, *47*, 1585–1593.
13. Ouh, H.K.; Choi, J.; Lee, J.; Han, S.; Kim, S.; Seo, J.; Ko, H. Capacitive Readout Circuit for Tri-axes Microaccelerometer with Sub-fF Offset Calibration. *J. Semicond. Tech. Sci.* **2014**, *14*, 83–91.
14. Lee, S.; Ko, H.; Choi, B.; Cho, D. Optimal and robust design method for two-chip out-of-plane microaccelerometers.*Sensors* **2010**, *10*, 10524–10544.

Chapter 4

Design of Ultra-Low Power PMOS and NMOS for Nano Scale VLSI Circuits

Ashok Babu Ch[1], J. V. R. Ravindra[2], K. Lalkishore[3]

[1]Department of Electronics and Communication Engineering, SVIT, Sec-Bad, India
[2]Department of Electronics and Communication Engineering, Vardhaman College of Engineering, Hyderabad, India
[3]Jawaharalal Nehru Technological University, Anantapur, India

ABSTRACT

CMOS devices play a major role in most of the digital design, since CMOS devices have larger density and consume less power. The integrated circuit performance mostly depends on the basic devices and its scaling methods, but in conventional CMOS devices in ultra deep submicron technology, leakage power becomes the major portion apart of dynamic power. The demerits of the conventional CMOS is less speed and, more leakage, for any digital design PDP is the figure of merit which can be used to determine energy consumed per switching event, hence we designed a NOVEL NMOS and PMOS which has superior performance than conventional PMOS and NMOS, the design and performance checked at 90 nm, 180 nm and 45 nm technology and calculate the performance values.

Keywords: Power Delay Product, Average Power, Static Power, Delay, Dynamic Threshold CMOS

1. INTRODUCTION

The need for low power chips is the increased marked demand for portable consumer electronics powered by the batteries, which have not experienced the similar rapid density growth compared to electronics circuits. It has been concluded that battery technology alone will not solve power problems in the near future. Therefore designing a chip with low power is becoming a major concern in the chip design industry. The different power reduction techniques from device level to circuit level and system level have been shown as a low power taxonomy [1].

In ultra-deep submicron technology especially below 45 nm technology, leakage power is becoming a major concern than the dynamic power. As the technology is going on scaling down, and we are now in 20 nm/15 nm/ 10 nm technology proper controlling of leakage power is becoming challenge factor.

Any power reduction techniques at device level will certainly lower the overall power consumption at system level (power reduction means it should be applicable from device level to system level, there is no guarantee that device level power reduction always applicable at system level) [2].

Related work: The existing numerous power reduction techniques in the field of VLSI are suitable for only 90 nm and above, but continuous scaling of the device in the present VLSI technology many of them may not be suitable. Moreover if the technique is suitable for power reduction than it fails for high speed, hence a new technique is required that is suitable for both power and speed [3].

Sources of power dissipations are

$$P_{\text{avg}} = P_{\text{static}} + P_{\text{dynamic}} + P_{\text{leakage}} + P_{\text{shortckt}}.$$

The major source in the overall power dissipation is the dynamic power

$$P_{\text{dynamic}} = \alpha \times V_{DD}^2 \times C_L \times f.$$

α = transition activity factor, V_{DD} = supply voltage, C_L = load capacitance, f = frequency of operation.

As the technology shrinks leakage power is to be dealt very carefully.

2. TAXONOMY OF LOW POWER

In Figure 1 different power dissipation and power reduction techniques are clearly shown, all these techniques are applicable in low power design [4] .

3. PROBLEM STATEMENT

Here an attempt has been made to control leakage power by proposed PMOS and NMOS, the fundamental limit of conventional PMOS and NMOS is higher leakage and lower sped in deep submicron technology sine power supply reduction lowers three times the threshold voltage will degrade the speed of the circuit considerably, as we are striving forward to higher level of integration, area is not having that much of concern, now we are in position to keep millions of transistors in a single die, the other performance parameters like power, speed and delay moreover power delay product are important parameters which can decide the role of the device in nano scale regime [5] . In design abstraction we are approaching bottom up design as shown in the below Figure 2, basically CMOS device level analysis is done and comparative study is done among different MOS devices.

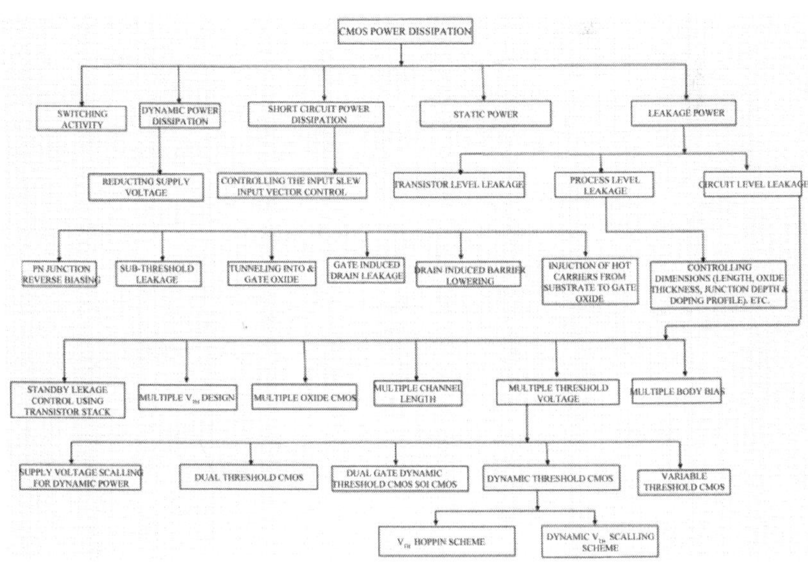

Figure 1. Taxonomy of low power.

Demerits of Conventional PMOS & NMOS

Figure 3(a) and Figure 3(b) show the conventional PMOS and NMOS and having the following demerits.

- Lower speed.
- Higher leakage in deep submicron technology.

Figure 4 in DT PMOS and DT NMOS body is connected to gate terminal of the MOS device and as the gate voltage is changing body biasing takes place.

Working of proposed PMOS and NMOS: in Figure 5 for top transistor (PM0) a reference voltage of 0.8 V is applied, and input is applied to the gate terminals of bottom transistor (PM2) and output is taken across the drain terminals of the right transistor (PM1) [6].

Both the conventional PMOS and NMOS, DT PMOS, DT NMOS and proposed PMOS and proposed NMOS are implemented using 45 nm technology and all are simulated using Cadence Virtuoso Design Environment and simulated and results are shown Table 1. From Table 1 the proposed PMOS and NMOS have low PDP value.

Comparison table for the conventional PMOS and NMOS, DT PMOS, DT NMOS and proposed PMOS and proposed NMOS are implemented using 90 nm technology and all are simulated using Cadence Virtuoso Design Environment and simulated and results are shown Table 2.

Comparison table for the conventional PMOS and NMOS, DT PMOS, DT NMOS and proposed PMOS and proposed NMOS are implemented using 180 nm technology and all are simulated using Cadence Virtuoso Design Environment and simulated and results are shown Table 3.

The PMOS and NMOS are good transfer of 1 and good transfer of 0, the comparative study is done on various types of MOS devices like PMOS, DTMOS and proposed MOS devices, the average power, delay and PDP is calculated by using the cadence virtuoso design environment tool for 45 nm, 90 nm, 180 nm technology [7] .

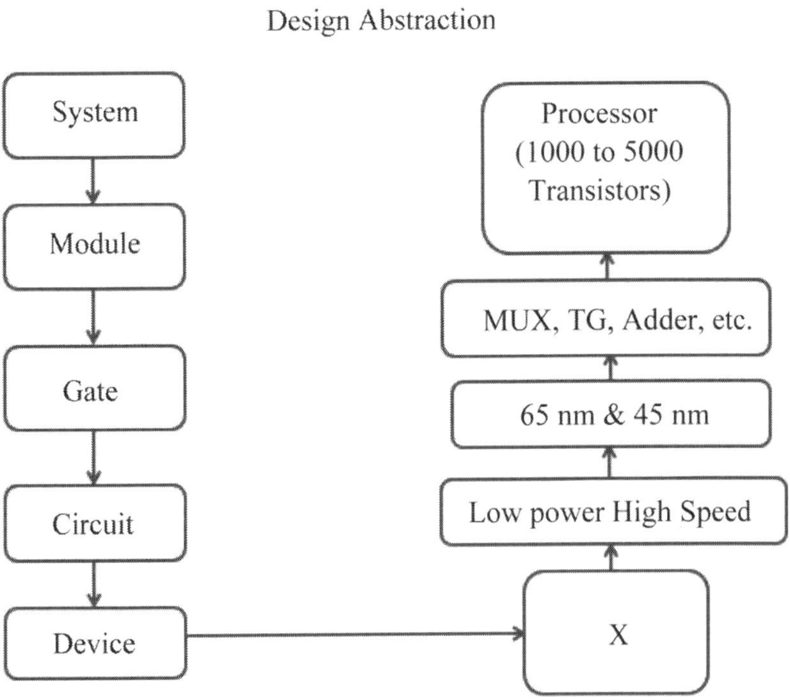

Figure 2. Design abstraction.

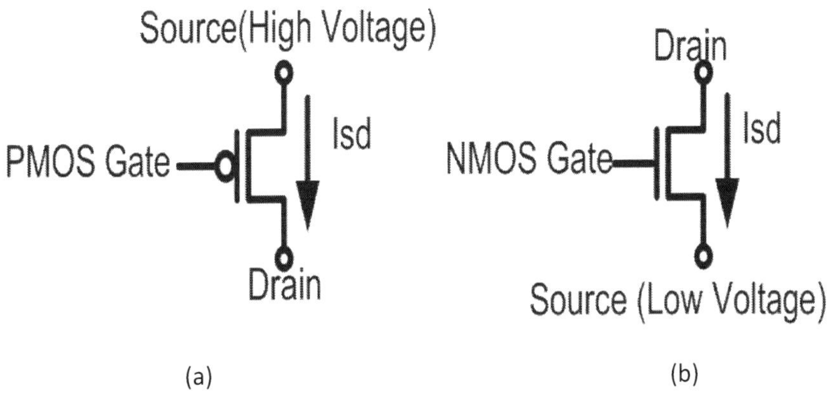

Figure 3. Conventional PMOS (a) & NMOS (b).

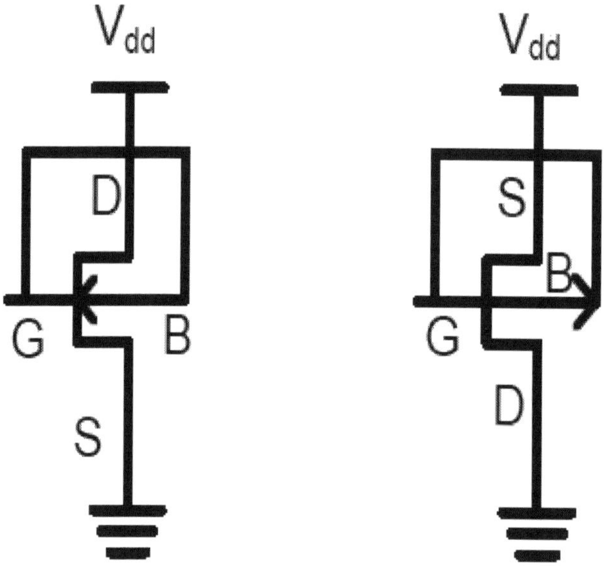

Figure 4. DT PMOS and DT NMOS.

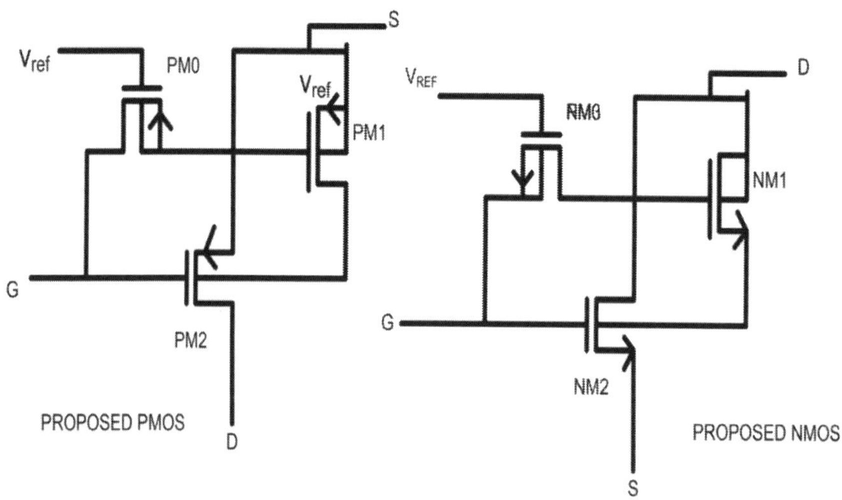

Figure 5. Proposed PMOS and NMOS.

Table 1. 45 technology.

Device	Output max value	Out min value	Avg. power	Delay	Static power		PDP
					$V_{IN} = 0$ V	$V_{IN} = 1.1$ V	
PMOS	1.1 V	0.323 V	5.37×10^{-9}	16 ps	116×19^{-15}	481×10^{-24}	8.59×10^{-20}
DTMOS	0.805 V	0.217 V	5.85×10^{-3}	4 ns	588×10^{-63}	11.26×10^{-3}	2.34×10^{-14}
NOVEL PMOS	1.1 V	0.250 V	1.71×10^{-9} V	4 ps	1.443×10^{-12}	822.5×10^{-15}	6.84×10^{-21}
NMOS	0.79	0	4.412×10^{-9}	13 ps	0	101.1×10^{-15}	5.73×10^{-20}
DTMOS	0.885 V	0.277 V	2.906×10^{-3}	4 ns	6.24×10^{-3}	48×10^{-27}	2.6154×10^{-14}
PROPOSED NMOS	0.818 V	0 V	6.29×10^{-9}	9 ps	151.8×10^{-15}	268.8×10^{-15}	5.661×10^{-20}

Table 2. 90 nm technology.

Device	Output max value	Out min value	Avg. power	Delay	Static power		PDP
					$V_{IN} = 0$ V	$V_{IN} = 1.1$ V	
PMOS	1.2 V	0.233 V	13.26×10^{-9}	5 ps	1.3×10^{-15}	774×10^{-27}	663×10^{-20}
DTMOS	0.618 V	0.142 V	10.59×10^{-3}	6 ps	275.7×10^{-45}	20.28×10^{-3}	6.35×10^{-14}
NOVEL PMOS	1.2 V	0.158 V	10 nw	3 ns	15.60×10^{-15}	3.052×10^{-12}	3×10^{-17}
NMOS	1.08 V	0 V	8.245×10^{-9}	2 ns	1.404×10^{-24}	1.306×10^{-15}	1.6×10^{-17}
DTMOS	1.15 V	0.55 V	17.18×10^{-3}	4 ns	37.32×10^{-3}	959.1×10^{-30}	6.87×10^{-11}
PROPOSED NMOS	1.16 V	0 V	5.39×10^{-9}	0.0092 ns	45.24×10^{-12}	3.12×10^{-15}	4.3×10^{-21}

Table 3. 180 nm technology.

Device	Output max value	Out min value	Avg. power	Delay	Static power		PDP
					$V_{IN} = 0$ V	$V_{IN} = 1.1$ V	
PMOS	1.8 V	0.674	1.086×10^{-6}	0.1215 ns	23.46×10^{-15}	32.74×10^{-30}	2.25×10^{-6}
DTMOS	0.6182	0.2942	991.40×10^{-3}	40 ps	0	1.8×10^{-3}	3.96×10^{-11}
NOVEL PMOS	1.8 V	0.450 V	1.08×10^{-6}	10 ps	46.96×10^{-15}	23.46×10^{-15}	108×10^{-17}
NMOS	1.22 V	0 V	0.398×10^{-6}	5×10^{-3}	522.8×10^{-48}	467.7×10^{-18}	199×10^{-9}
DTMOS	1.54 V	1 V	37.69×10^{-3}	10 ns	81.7×10^{-3}	106.2×10^{-30}	3.769×10^{-10}
PROPOSED NMOS	1.435 V	0 V	1.08×10^{-6}	6 ps	487.7×10^{-18}	$935.3 \times ^{10-18}$	648×10^{-18}

4. CONVENTIONAL CMOS INVERTER

Figure 6 shows a conventional CMOS inverter which is simulated using Cadence Virtuoso Design tool and simulated outputs are shown in Figure 7, Figure 8 shows CMOS output when all Zero's Applied and Figure 9 shows CMOS output when All one's Applied.

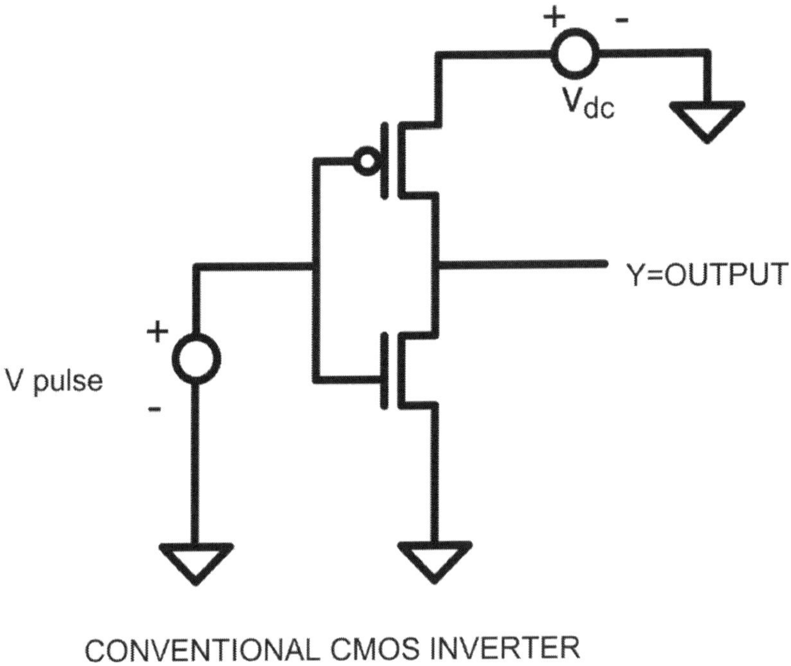

Figure 6. Conventional CMOS inverter.

5. INVERTER WITH PROPOSED PMOS AND NMOS

From the Figure 10 for T_6 and T_3 applied a reference voltage of 0.8 V when V_{in} = 0 V applied $T_4$and T_5 become on and output node will be 1 and T_1, T_2 will become off. When input is 1 T_4, T_5will be non conducting and T_2, T_1 will become conducting will be zero as current sinks towards 0 v [8]. And output results are shown in the Figures 11-13.

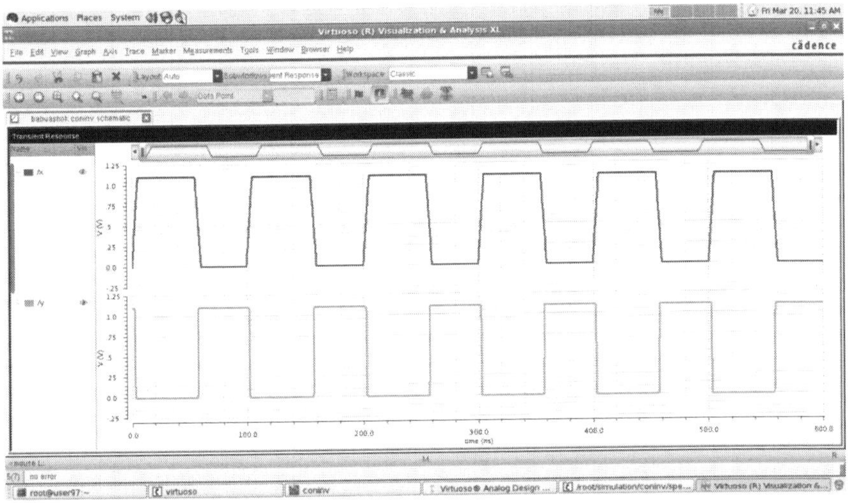

Figure 7. Conventional CMOS inverter output.

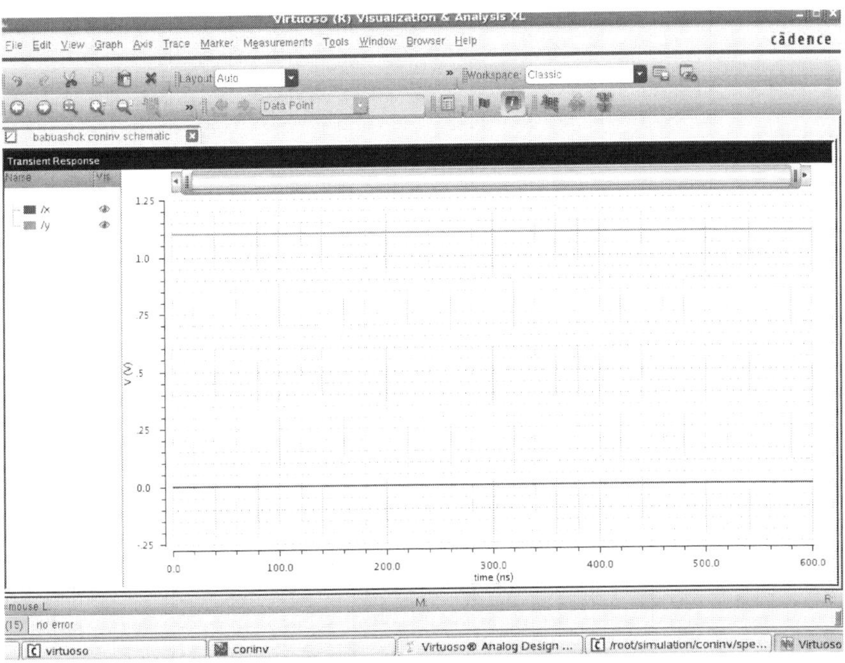

Figure 8. CMOS output when all zero's applied.

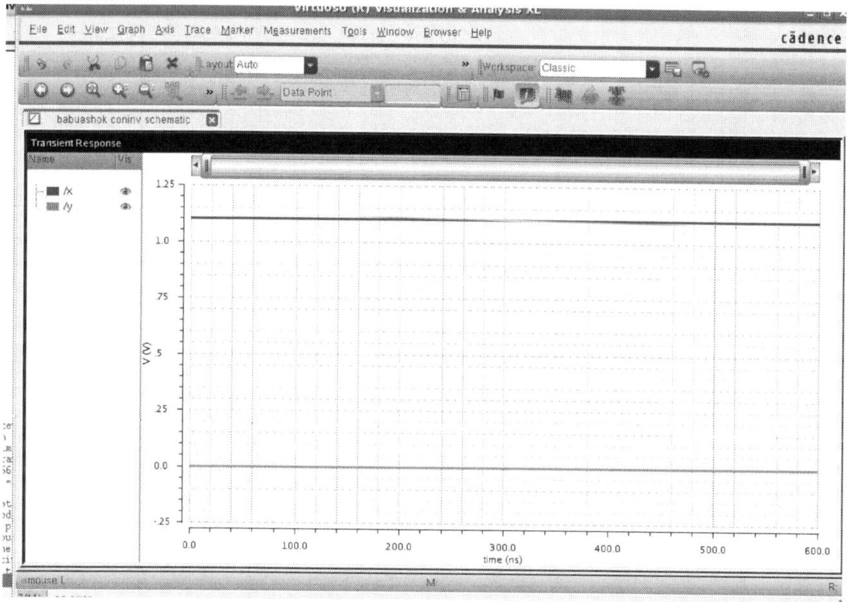

Figure 9. CMOS output when all one's applied.

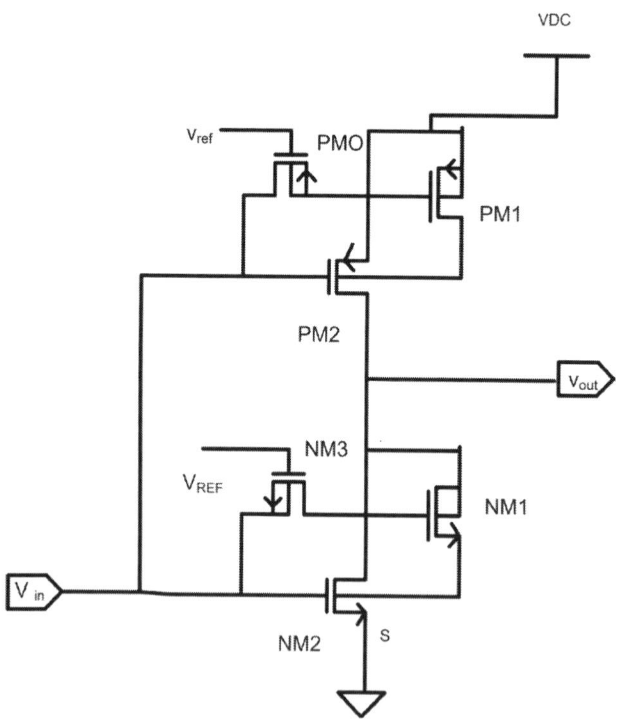

Figure 10. Proposed CMOS inverter.

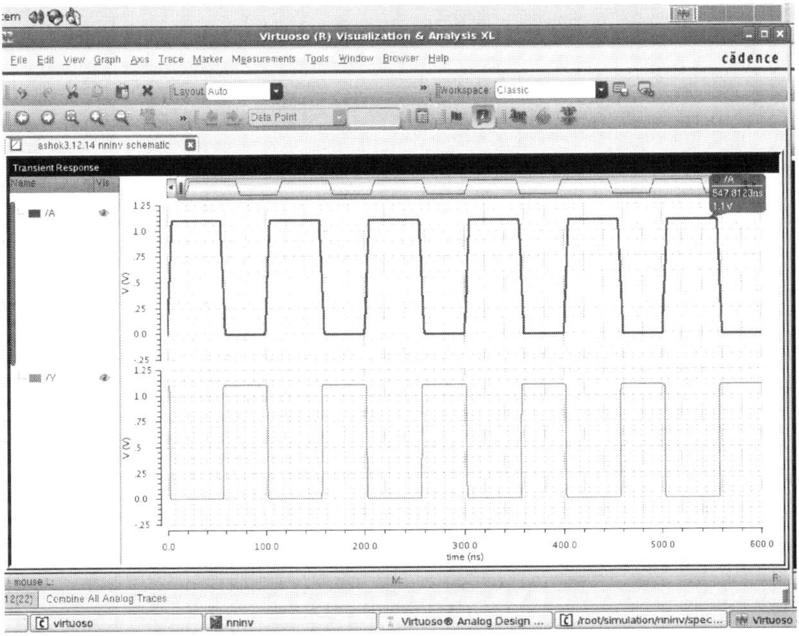

Figure 11. Proposed CMOS inverter output.

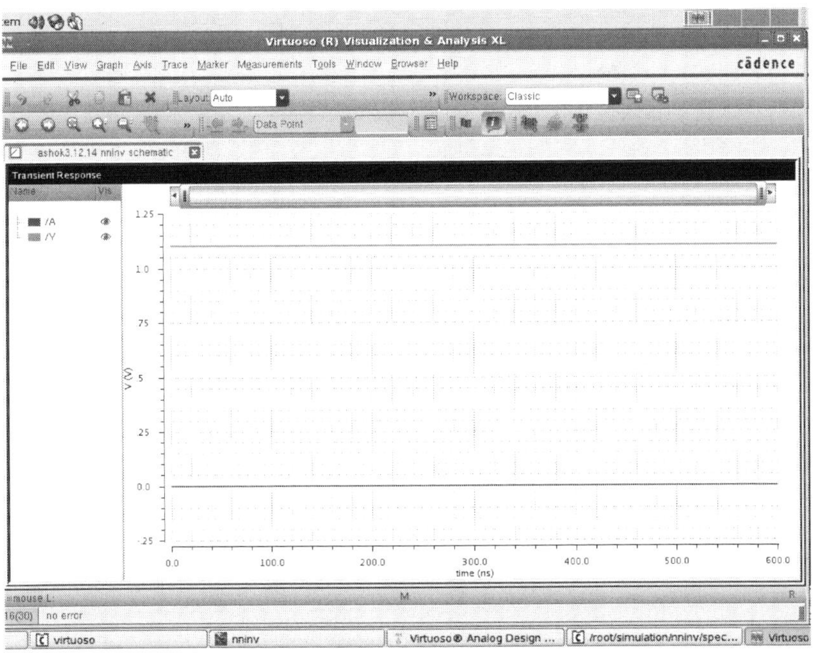

Figure 12. Proposed CMOS output for all zero's.

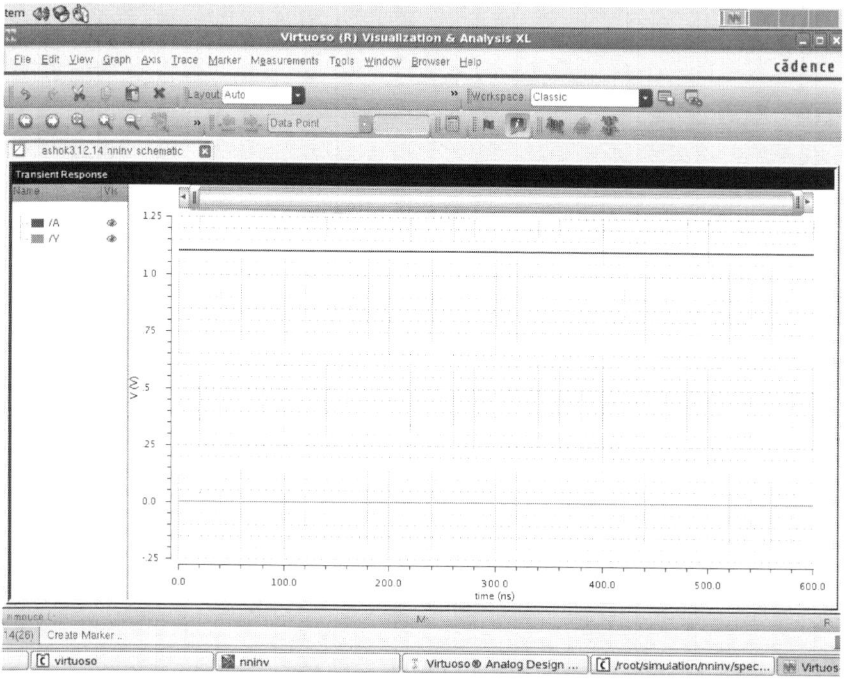

Figure 13. Proposed CMOS output for all one's.

There may be an area overhead but this circuit is certainly suitable for high speed and low leakage in deep submicron technology, as area is not a problem since we are in a position to keep millions of transistors in single die, performance parameters like speed will increase and leakage will decrease [9].

Hence an attempt has been made with proposed MOS devices to improve the performances.

6. LEAKAGE CURRENT MECHANISM

Reverse biased junction leakage and subthreshold leakage have very much similar characteristics both will be in order of pico amperes per device and very sensitive to process variations. Leakage current cannot be ignored in deep submicron region and leakage current is the beyond the digital designer control [10] . In large scale high performance digital chips performing operations with high frequency. In ultra submicron region leakage power is playing a major contribution apart of dynamic power [11].

Subthreshold channel leakage: the subthreshold conduction current is given by $I_{sub} = I_0 e^{v_{gs} - v_t / \alpha v_{th}}$ where alpha is parameter depends on fabrication process v_{th} is the threshold voltage, v_{th} is the thermal voltage, even though a transistor is logically turned off there may be non zero leakage current through the channel this current is known as Subthreshold leakage current Since it occurs below threshold voltage [12] .

7. CALCULATION OF LEAKAGE POWER IN CMOS INVERTER AND PROPOSED CMOS INVERTER AT 45 NM TECHNOLOGY

7.1. Normal CMOS Inverter Leakage Power

- Power: 6.937 ´ 10^{-9}, delay: 0.05 ´ 10^{-9}.
- When all ones applied power is 4.917 ´ 10^{-12}.
- When all zeros applied power is 5.018 ´ 10^{-12}.
- Total power for all ones and all zeros is 4.97 ´ 10^{-12} + 5.018 ´ 10^{-12} = 9.935 ´ 10^{-12}.
- Leakage power = 6.937 ´ 10^{-9} – 9.935 ´ 10^{-12} = 6927 ´ 10^{-12}.

7.2. Novel CMOS Inverter Leakage Power

- AVG power P = 6.849 ´ 10^{-9}, delay = 0.18 ns.
- Power when all ones applied = 6.655 ´ 10^{-12}.
- Power when all zeros applied = 11.44 ´ 10^{-12}.
- All ones + all zeros power = 6.655 ´ 10^{-12} + 11.44 ´ 10^{-12} = 1.8095 ´ 10^{-11}.
- Leakage power = 6.849 ´ 10^{-9} – 1.8095 ´ 10^{-11} = 6830 ´ 10^{-12} w.
- The difference of leakage between conventional CMOS and Proposed CMOS power is
- 6927 ´ 10^{-12} – 6830 ´ 10^{-12} = 97 ´ 10^{-12}.

8. CONCLUSION

We designed a Novel PMOS and NMOS which could give superior performance in deep submicron technology. The MOS devices like PMOS, DTMOS, and Novel MOS are studied and PDP is comparatively calculated. The leakage power is calculated for normal CMOS inverter and Novel CMOS inverter and it is observed that Novel CMOS inverter has less leakage power compared to Normal CMOS inverter. There may be area overhead but performance is better. It is concluded that Novel PMOS and NMOS are certainly suitable for low leakage and low power in ultra-deep submicron technology [13].

REFERENCES

1. Burch, R., Najm, F., Yang, P. and Trick, T. (1993) A Monte Carlo Approach for Power Estimation. IEEE Transactions on Very Large Scale Integration (VLSI) Systems, 1, 63-71.
2. Ghosh, S. Devadas, K Keutzer, and J. White, (1992) Estimation of Average Switching Activity in Combinational and Sequential Circuits. ACM IEEE Design Automation Conference, 253-259.
3. Chou, T.L. and Roy, K. Statistical Estimation of Digital Circuit Activity Considering Uncertainty of Gate Delays. IEICE (Japan) Transactions on Fundamentals of Electronics, Communications and Computer Sciences, Accepted for Publication.
4. Chou, T.-L. and Roy, K. (1996) Accurate Estimation of Power Dissipation in CMOS Sequential Circuits. IEEE Transactions on VLSI Systems, 369-380.
5. Kumar, M., Hussain, Md.A. and Paul, S.K. (2013) An Improved SOI CMOS Technology Based Circuit Technique for Effective Reduction of Standby Subthreshold Leakage. Circuits and Systems, 4, 431-437.
6. Kado, Y. (1997) The Potential of Ultrathin-Film SOI Devices for Low-Power and High-Speed Applications. IEICE Transactions on Electronics, E80-C, 443-454.
7. Cristoloveanu, S. and Reichert, G. (1998) Recent Advances in SOI Materials and Device Technologies for High Temperature. Proceedings of High-Temperature Electronic Materials, Devices and Sensors, San Diego, 22-27 February 1998, 86-93.
8. Brglez, F. and Fujiwara, H. (1985) A Neutral Netlist of 10 Combinational Benchmark Circuits and a Target Translator in Fortran. IEEE International Symposium on Circuits and Systems.

9. Chew, E.S., Phyu, M.W. and Goh, W.L. (2009) Ultra Low-Power Full-Adder for Biomedical Applications. IEEE International Conference of Electron Devices and Solid-State Circuits, Xi'an, 25-27 December 2009, 115-118.
10. Shalem, R., John, E. and John, L.K. (1999) A Novel Low Power Energy Recovery Full Adder Cell.
11. Chen, Z., Roy, K. and Chou, T.-L. (1997) Sensitivity of Power Dissipation to Uncertainties in Primary Input Specification. IEEE Custom Integrated Circuits Conference, 487-490.
12. Chen, Z., Roy, K. and Chou, T.-L. (1997) Power Sensitivity—A New Method to Estimate Power Considering Uncertain Specifications of Primary Inputs. ACM/IEEE International Conference on Computer-Aided Design, 40-44.
13. Najm, F. (1994) A Survey of Power Estimation Techniques in VLSI Circuits. IEEE Transactions on Very Large Scale Integration (VLSI) Systems, 2, 446-455.

Chapter 5

Active C4 Electrodes for Local Field Potential Recording Applications

Lu Wang [1,*], David Freedman [1], Mesut Sahin [2], M. Selim Ünlü [1,3] and Ronald Knepper [1]

[1]Department of Electrical and Computer Engineering, Boston University, 8 Saint Mary's St, Boston 02215, MA, USA
[2]Department of Biomedical Engineering, New Jersey Institute of Technology, 323 Martin Luther King, Jr. Boulevard, University Heights Newark, Newark 07102, NJ, USA
[3]Department of Biomedical Engineering, Boston University, 44 Cummington St, Boston 02215, MA, USA

ABSTRACT

Extracellular neural recording, with multi-electrode arrays (MEAs), is a powerful method used to study neural function at the network level. However, in a high density array, it can be costly and time consuming to integrate the active circuit with the expensive electrodes. In this paper, we present a 4 mm × 4 mm neural recording integrated circuit (IC) chip, utilizing IBM C4 bumps as recording electrodes, which enable a seamless active chip and electrode integration. The IC chip was designed and fabricated in a 0.13 μm BiCMOS process for both *in vitro* and *in vivo* applications. It has an input-referred noise of 4.6 μVrms for the bandwidth of 10 Hz to 10 kHz and a power dissipation of 11.25 mW at 2.5 V, or 43.9 μW per input channel. This prototype is scalable for implementing larger number and higher density electrode arrays. To validate the functionality of the chip, electrical testing results and acute *in vivo* recordings from a rat barrel cortex are presented.

Keywords: 3D electrodes; C4; CMOS; extracellular; in vitro; in vivo; MEA; neural sensors

1. INTRODUCTION

Extracellular recordings of neural activity with an array of electrodes placed on the brain surface or juxtaposed to an extracted brain slice have been shown to be very useful for studying neural function at the network level [1,2]. To achieve high spatial resolution, high density multi-electrode arrays (MEAs) are often used for extracellular recordings [3,4,5]. These MEAs can record the low frequency field potentials generated by local neurons, and sometimes high frequency neural spikes if they are sufficiently close to individual cells, even though the extracellularly recorded neural signals are on the order of tens of microvolts. Both local field potentials (LFP) at the lower end of the frequency spectrum and the multi-unit activity in the kilohertz region contain functional information [6,7]. There is an increasing demand for larger count electrodes with increased spatial density in order to collect the maximum amount of information from the neural tissue of interest. Large amplification and low input-referred noise are also required due to the low amplitudes of extracellular potentials. Additionally, power density must be kept below 0.8 mW/mm2 to prevent damage to the surrounding neural cells [8]; this requirement demands minimal power dissipation from the system [9].

High density active MEAs with small chip areas have been developed to achieve high spatial resolution [10,11,12]. Usually a large amount of post processing is required to fabricate the 2D electrodes on top of a standard CMOS chip. Passive MEAs with 3D electrodes, such as the Utah electrode array (UEA) [8,13,14,15,16,17], and flexible or rigid multi-shank electrode arrays [18,19,20,21], have been integrated with CMOS integrated circuits (ICs) to create active MEAs. A capacitive feedback pre-amplifier topology [22] has been widely adopted in the neural recording system design [15,23,24,25]. The voltage gain of this pre-amplifier is determined by the ratio of its input coupling capacitor and the feedback capacitor. However, the use of this approach also limits its scalability, since the capacitance per unit area does not increase linearly as CMOS technology scales down. Critically, commercially available UEAs have a maximum of 100 electrodes with a minimum pitch of 400 μm, which further limits the scalability of such active MEAs.

In this paper, we present the characterization and preliminary animal testing results of a scalable, CMOS compatible, neuropotential recording IC with C4 (controlled collapsed chip connect) solder bumps as 3D electrodes, as shown in Figure 1. The recording system has AC-coupled single-ended input channels with PMOS transistors biased in the accumulation region as pseudo-resistors to achieve sub-Hertz low-frequency bandwidth, and uses the cascode transistor of the input stage as a switch, which enables input channel multiplexing for data serialization and tail current reuse for power reduction. Current conveyor circuitry is incorporated into the design to achieve high voltage gain with improved stability. This is the first time C4s have been demonstrated as electrodes for *in vivo* recording.

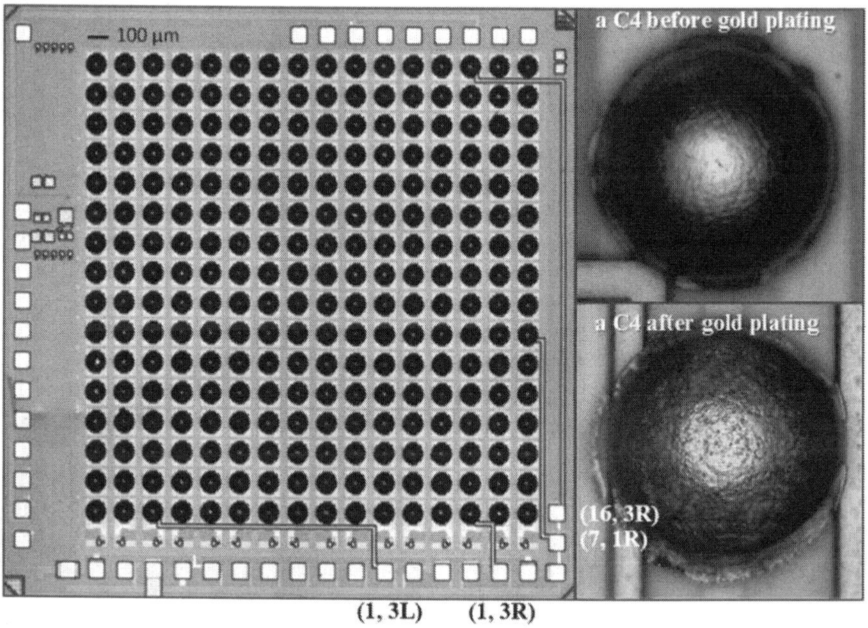

Figure 1. Micrograph of the fabricated neural recording integrated circuit (IC). The input channels with direct connections to the package are all labeled near the chip's corresponding pads. The C4 image is compiled from a number of different focus points on the microscope; *i.e.*, to see the whole C4, this and the previous C4 images were assembled from multiple Z focus-positions.

2. CHIP DESIGN

The architecture of the neuropotential recording IC chip is shown in Figure 2. The active MEA has 256 on-chip C4 electrodes for extracellular recording, occupying an area of 3.2 mm × 3.2 mm (active area), or 25 electrodes per mm2. Each C4 electrode is capacitively-coupled to a low-noise pre-amplifier to form an input channel. The AC coupling can prevent the system from becoming saturated by the half-cell potential of the electrode contacts. Sixteen rows are selected sequentially by an on-chip counter, and two analog multiplexers are used to select one of eight columns on both the left and right sides. At any given time, only one column on each side of the chip (left and right) is connected to an output channel, and thus only two output channels are needed for the 256 input channels. In the following sections, an input channel row x, column y on the left side will be referred as input (x, yL). Testing was performed on a custom-designed printed circuit board (PCB).

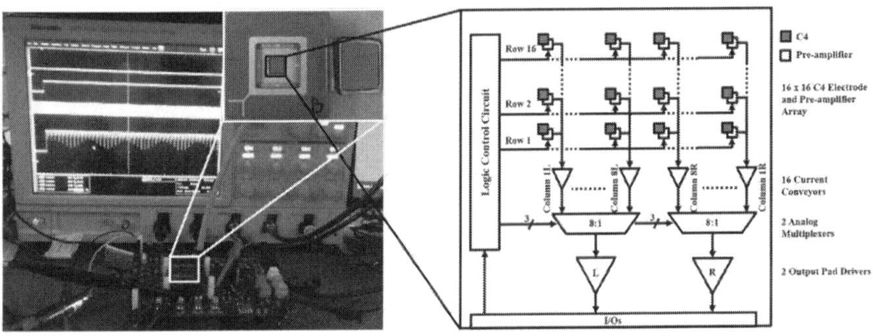

Figure 2. Neuropotential acquisition system and block diagram of the neuropotential acquisition IC chip architecture. For the prototype, the chip was designed with two analog outputs, and the digitization and signal processing are performed on the board level to reduce the design complexity.

2.1. C4 Electrodes

C4s are standard IBM solder balls used in flip-chip bonding, with different sizes and pitches available. The smallest size available at the time of this chip fabrication is "3 on 6" (mil), which has a diameter of 75 μm and a pitch of 150 μm [26]. This chip is designed using a 130 nm technology that comes with the C4 option of "4 on 8", so each C4 electrode occupies a chip area of 0.04 mm2. This area is much larger compared to individual pre-amplifiers and even the metal-insulator-metal (MIM) capacitors, which offer about 1fF per μm2 in this technology, so the C4 or any other electrode is often the major limiting factor of the MEA density. For example, with "2 on 4" C4s [26], the same chip area will be able to accommodate 1024 electrodes. Continued development is increasing pitch densities, *i.e.*, the most advanced fine pitch Cu pillar bumps have a staggered pitch of 30/60 μm [27].

An advantage of using C4s as on chip electrodes is that they are compatible with the standard CMOS process, and minimum post processing is required, which leads to low cost and high yield. Also, for single unit recording, it is possible to integrate penetrating electrodes with the presented IC by flip-chip bonding [15].

The equivalent circuit model used for a C4 electrode in this work is shown in Figure 3 and was adopted from [28,29,30]. Ri and Ci represent the electrode interface resistance and capacitance, respectively. Rs models the spreading resistance. Ri models the charge transfer resistance and has been theoretically determined to be given by the low-field approximation to the Butler-Volmer equation [28]. Ci is derived from the determination of the constant phase angle impedance ZCPA [28]. However, in this work, both Ri and Ci were determined from measurements of C4 impedance, which will be reported later in Section 4.3.

The spreading resistance Rs will depend to some extent upon the position of the C4 on the 16 × 16 array. For this work we use a calculated value of 3.18 KΩ, determined from the expression Rs=ρs/2πrei, where the resistivity of the solution ρs=100Ω−cm, and the radius rei of the C4 electrode is 50 μm.

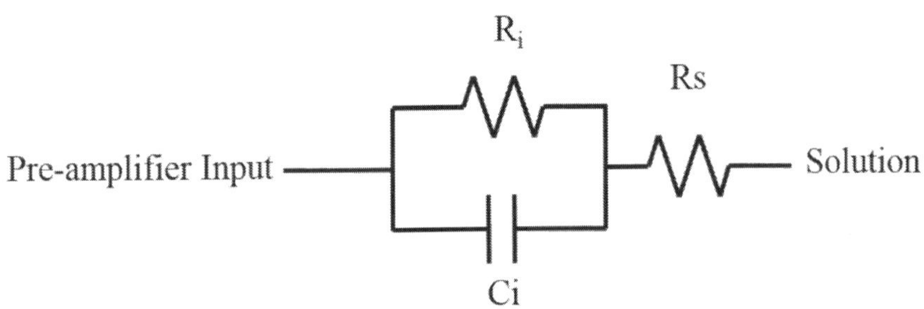

Figure 3. Equivalent circuit model for C4 electrode.

2.2. Low-Noise Preamplifier

The circuit schematic of the pre-amplifier array is shown in Figure 4. Each pre-amplifier from the array is a single-ended input PMOS cascode amplifier. The input of each pre-amplifier is AC coupled to the C4 recording electrode through a 10 pF MIM capacitor Cin. The input of the pre-amplifier is also DC coupled to a fixed voltage V1 (*i.e.*, 1.8 V) through a high-impedance pseudo-resistor M0 to provide biasing. The PMOS pseudo-resistor can achieve a GΩ-TΩ resistance value R0 with transistor M0's gate biased in the sub-threshold region [18,22], or even in the accumulation region [31]. The low frequency corner fL of the pre-amplifier is determined by the dominant capacitor Cin and the pseudo-resistor R0. Its equation is given by

$$f_L = \frac{1}{2\pi C_{in} R_0} \tag{1}$$

The function of the calibration cell will be discussed in Section 2.4.

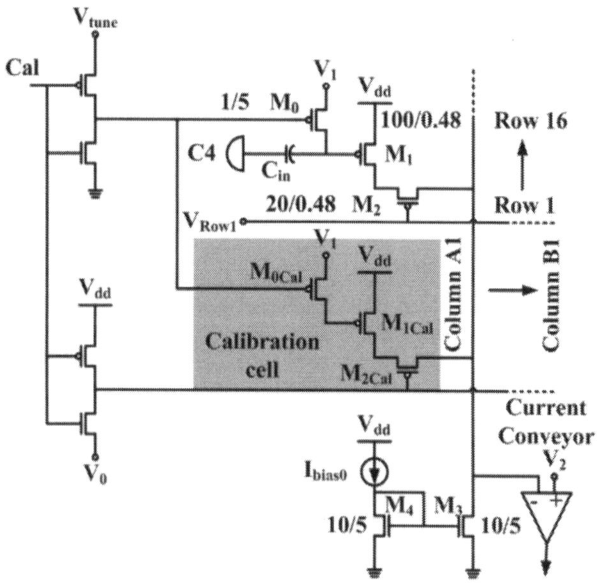

Figure 4. Circuit schematic of the pre-amplifier array. The calibration cell has the same size as a regular pre-amplifier, only without being connected to a C4 input electrode. There is one calibration cell on each column. The substrates of NMOS and PMOS devices are connected to the ground and the 2.5 V power supply, respectively.

To achieve the desired fL value, a larger R0 allows the use of a smaller input coupling capacitor. Theoretically, to the first-order approximation, with a TΩ resistor, a 160 fF capacitor is enough to achieve a 1 Hz low frequency corner. A single-ended input topology with a high resistance pseudo-resistor doesn't rely on large value capacitors, and thus allows this approach to take advantage of further scaling.

The input-referred thermal noise voltage of the pre-amplifier is:

$$\overline{v_n^2} = \frac{8kT}{3g_{m1}} \cdot (1 + 2 \cdot \frac{g_{m3}}{g_{m1}}) \cdot \Delta f \tag{2}$$

Therefore, to reduce the thermal noise, a large gm1 and a small gm3 is desired. In terms of transistor size, a larger gate width over a length ratio of M1 (100 µm / 0.48 µm) was used, compared to M3 (10 µm/5 µm). There is a trade-off between minimizing the thermal noise and the 1/f noise, since the latter requires a small gm1 value.

The input-referred root mean square (rms) noise voltage is calculated using the noise frequency curve by

$$\overline{v_{rms}} = \sqrt{\int_{10Hz}^{10kHz} \frac{\overline{v_n^2}}{\Delta f} \mathrm{d}f} \tag{3}$$

which gives 3.99 µVrms.

2.3. Current Conveyor

The output of the pre-amplifier is connected to the current conveyor. As shown in Figure 5, the current conveyor is a current amplifier followed by a transimpedance structure (M16–M19) and an NMOS source follower. With the inverting input of the differential stage connected to a feedback loop, the DC bias voltage follows the non-inverting input V2 that is supplied from an on-chip voltage reference. This is the primary reason a current conveyor is used here because the voltage of the inverting input is the drain voltage of transistor M3 of the pre-amplifier array. As shown in Figure 4, all 16 pre-amplifiers in a column share the same transistor M3, so any change in the drain voltage of M3 will affect the performance of the whole column. The current conveyor stage is essentially a transimpedance amplifier, and its gain is heavily dependent on the output impedance of transistor M17 and M19, which convert the current signal back to the voltage domain, so a transistor length of 2.5 μm is used for both M17 and M19. The total voltage gain Av, on the first order, is

$$A_v = g_{m1} \times (r_{o17} || r_{o19}) \tag{4}$$

where gm1 is the transconductance of transistor M1 in Figure 4. The total gain of the first two stages is simply the input transistor transconductance times the output impedance, because the current conveyor mirrors the signal current with a ratio of 1:1.

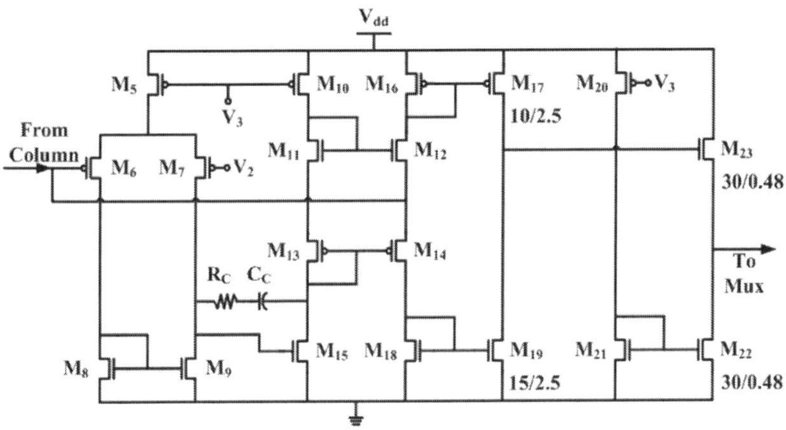

Figure 5. Circuit Schematic of the current conveyor amplifier.

The signals generated by the logic control circuit are non-overlapping, which means there will be a short time when all the pre-amplifiers in a column are shut down. The existence of the current conveyor will provide a DC path to Vdd for the drain of M3 during this time. This path will prevent the drain of M3 from being pulled down to ground. Otherwise, switching between rows would cause severe voltage spikes on the output waveform and most importantly would slow down the frame readout speed of the neural recording chip.

To ensure stability, a dominant-pole compensation capacitor Cc is used to control the dominant first pole and to split it from the high frequency poles [32] so that the open-loop gain of the amplifier's transfer function will be brought down to unity before the frequency reaches the high frequency poles. Also, the resistor Rc is included to introduce a left-half-plane zero above the unity gain frequency in order to relax the high frequency poles' effect on the phase. A simulated phase margin of 86.65° was realized with this RC compensation technique.

2.4. Time-Division Multiplexing and Chip Operation Modes

For high density MEAs, time-division multiplexing (TDM) [18] and sample-and-hold (SH) scheme [15] are often used to reduce the total number of output channels. Additionally, further digital compressive modulation schemes are often required for wireless systems because of their restrained bandwidth [33,34,35]. In this paper, for simplicity, TDM is utilized, so 256 signals recorded by the input amplifiers will be sequentially multiplexed onto 2 output channels to minimize the power consumption, chip area, and the number of chip I/Os. As shown inFigure 6a, a synchronous logic control circuit is used to carry out the TDM scheme. Every time when the chip is powered on, the low enabled $\overline{\text{reset}}$ signal is asserted to perform a power-on reset (POR). During POR, $\overline{\text{cal}}$ and $\overline{\text{stop}}$ pins are kept high, the output of the "1" hot counter will be reset to its initial value 1, so VRow1 is set to V0 (around 0.9 V) (On-chip current mirror bias circuits establish voltages V0–V3 and Vcal through off-chip current sources that have nominal values of 100 μA. The tuning range of the current source is 0 to 200 μA. Vtuneis normally set at 2.5 V.), whereas VRow2 to VRow16 all equal Vdd. VRow1 to VRow16 are used to control the gate of the cascode transistors M2, as shown in Figure 4, and thus during POR (or reset mode in general), only row 1 will be selected. Similarly, during POR, the output of the divider chain is reset to 111, and the 8-to-1 multiplexer connects column 1 to the output channel, so one column is selected from each side of the chip.

After POR, the $\overline{\text{reset}}$ pin is deasserted, the output of the divider-chain will count from 111 to 000 every 8 clock cycles, whereas the output of the counter repeats from 1 to 16 every 128 clock cycles. Thus, a consecutive input channel is selected every clock cycle, from input (1, 1) to (16, 8). Since at any given time only one of the M2transistors in a column is selected, all the pre-amplifiers in a column can share the same tail-current. As a result, the power consumption of the pre-amplifier array is reduced by a factor of 16. This is the continuous mode. The clock is designed to operate at 2.56 MHz in continuous mode, so each input channel will be selected every 128 clock cycles, for a 20 kHz sampling rate per channel (The design scan rate of 2.56 MHz has not been achieved in testing due to the impact of process variation. This will be discussed in Section 4).

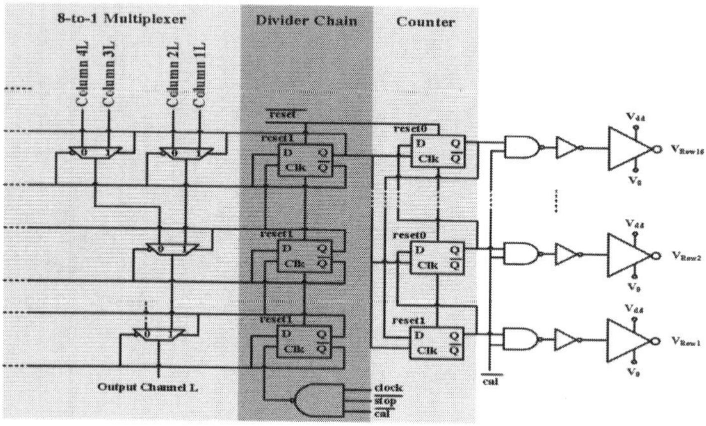

(a) Simplified circuit schematic.

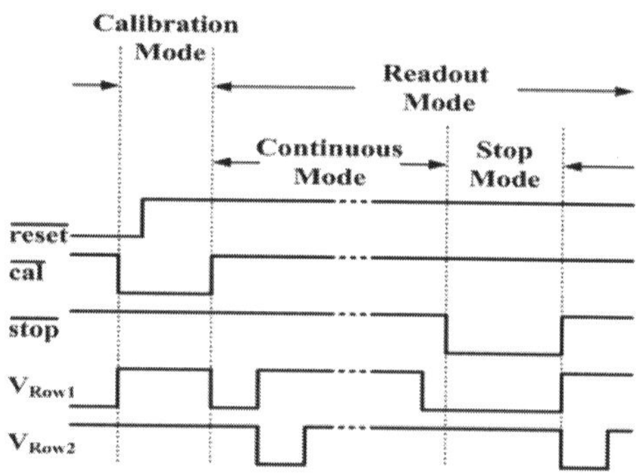

(b) Timing diagram.

Figure 6. The logic control circuit. (**a**) The schematic shows only half of the 8-to-1 multiplexer. The $\overline{\text{reset}}$ signal is routed through some of the D-latches simply to reduce clutter in the figure; (**b**) The timing diagram is an illustration of the chip's operation modes; the clock and all the control signals $\overline{\text{reset}}$, $\overline{\text{cal}}$ and $\overline{\text{stop}}$ are provided off-chip.

The chip can also operate in two additional modes: stop mode and calibration mode. Pulling down the $\overline{\text{stop}}$ pin will enable the stop mode, whereby the clock will be superseded, interrupting the TDM. Thus, stop mode allows observation in real time of any channel from each half of the chip. Continuous mode is resumed once the $\overline{\mathrm{stop}}$ signal is disabled. After every reset, a calibration often needs to be performed. As mentioned previously, the pseudo-resistor enables the recording of LFPs with small input capacitors. However, this high resistance value results in a relatively large RC time constant, which slows down the process of charging and discharging the gate of the M1 transistor. In order to accelerate this process, the $\overline{\mathrm{cal}}$ signal is used to turn on M0 to convert the pseudo-resistor into a low resistance state to acquire a desirable RC time constant that can calibrate the gate voltage of M1 to V1 efficiently. During calibration, the gate of M2Cal (as shown in Figure 4) will be pulled down to V0 to activate the calibration pre-amplifiers, while all the regular pre-amplifiers are disabled. The purpose of the calibration pre-amplifiers is to avoid the drain voltage of M3 being pulled down to ground during calibration. By keeping the drain voltage of M3 stable, it is possible to increase the switching speed between operation modes. The timing diagram of the operation modes is plotted in Figure 6b.

3. FABRICATION AND PACKAGING

The neural recording chip was designed and fabricated in IBM 0.13 μm BiCMOS 8HP technology (No bipolar devices were used in the chip in order to keep the design CMOS compatible). It has a total area of 4 mm × 4 mm and a recording area (active area) of 3.2 mm × 3.2 mm. A micrograph of the chip is shown in Figure 1. The C4 electrodes are standard "4 on 8" (mil), 100 μm in diameter with a 200 μm pitch. Their material composition is 97% Pb (lead) and 3% Sn (tin) (Lead free version of C4s are also available with 97.7% Sn and 2.3% Ag (silver). Scheduling issues necessitated using Pb-based C4s on this chip). To improve the C4 electrode biocompatibility, gold was plated onto the C4s using Bright Electroless Gold solution (Transene Company, Inc.) [36,37]. The photographs of C4s before and after Au electrolessplating are shown in Figure 1.

In order to accommodate the different requirements of electrical, *in vitro*, and *in vivo* testing, the fabricated chips were wire bonded to three different packages (Quik-Pak) [38], as shown in Figure 7. For electrical testing, a 10 × 10 ceramic PGA package was used to provide direct access to input channels (1, 3L), (1, 3R), (7, 1R), and (16, 3R) that were pre-wired to the chip pads with the top level metal, as shown in Figure 1. For *in vitro* applications, the neural recording chip was packaged in an open-top PGA package with bonding wires encapsulated for mechanical and electrical protection. The finished package has all the C4 electrodes exposed in order to interface with the brain/tissue slices, and a plastic fluidic chamber was built on top of the PGA package. An important function of the *in vitro* testing package is to provide a way to apply signals to all 256 input channels.

A 6 mm × 7 mm custom PCB package was also built to carry out acute *in vivo* experiments in living rats. The diameter of a rat brain cross-section is not much bigger than 1 cm, so the major requirement of the *in vivo* package is that it has to be able to fit on top of the rat brain. Meanwhile, it has to be connected to the recording side with a minimum number of wire connections without completely constraining the rat under test. In order to do so, the chip was wire bonded to the top side of the PCB with the bond wires encapsulated. The height of the epoxy was milled down to 100 μm after the encapsulation for better contact between the electrodes and rat brain. A 0.3 mm pitch, 17 pin flat flexible cable (FFC) was chosen as the interconnection between the *in vivo* package and a recording PCB. The width of the FFC was 5.1 mm, which is slightly smaller than the width of the PCB package, and is thin and flexible, all of which makes the FFC suitable for acute *in vivo* testing.

Figure 7. Chip packaging: A standard 10 × 10 ceramic PGA (**Top left**), an open-top 10 × 10 ceramic PGA (**Top middle**), front side of the custom printed circuit board (PCB) package (**Top right**), back side of the custom PCB package (**Bottom right**), and flat flexible cables (FFC) (**Bottom middle**).

4. CHIP CHARACTERIZATION

4.1. Electrical Testing

4.1.1. System Voltage Gain

The electrical testing package was used to characterize the voltage gain of a single input channel. Figure 8shows the measurement when the IC is operating in continuous mode at the maximum frame rate of 20 kHz. 128 inputs (per side) were scanned with only one input channel connected to the function generator. A sinusoidal signal with a peak-to-peak amplitude of 1 mV and a frequency of 1 kHz was applied to input channel (1, 3L) through an subminiature version A (SMA) connector. All the other input channels were left floating. The sine wave at the output can be seen in the highlighted envelope of the selected cell with the provided input. Calculation using the amplitude of the waveform envelope gives a voltage gain of 58.1 dB.

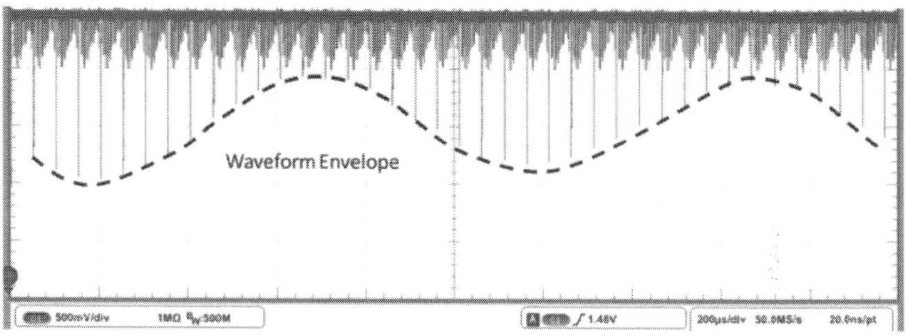

Figure 8. Measured output waveform from the recording chip in continuous mode with a sampling rate of 20 kHz. The envelope of a single channel that had a 1 kHz sinusoidal input is highlighted with the dashed line.

All 128 input channels on the left side of the chip were measured in stop mode using the *in vitro* testing package. The output DC voltage level of each input channel was manually adjusted to 1.5 V. The mean and standard deviation of the voltage gain of the 128 inputs on the left side of the chip were 58.7 dB and 0.37 dB, respectively.

4.1.2. Process Variation

Semiconductor process variation is the deviation of the manufactured device or interconnect parameters from their designed or expected nominal values. It can cause mismatch in current mirrors, different input recording channels, and threshold voltage variations. As a result, the chip output DC level will vary from channel to channel. In order for the neuropotential recording chip to work properly, the output DC level needs to be corrected for each individual input channel. This can be done by adjusting the off-chip current sources as shown in Figure 9. However, to operate the chip in continuous mode, the variation correction process needs to be integrated into the recording PCB so that during every clock cycle a proper bias current will be applied to the chip to automatically calibrate the DC voltage value of the output channel.

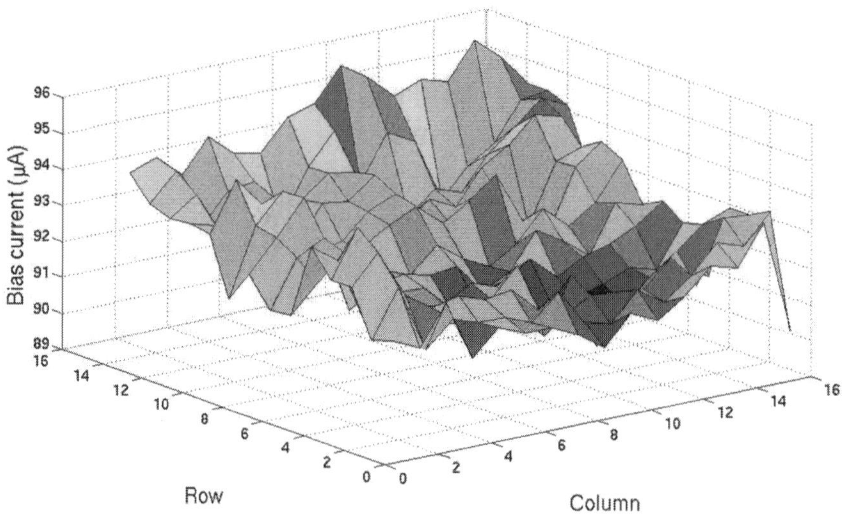

Figure 9. Bias current values to set the output DC level of each individual input channel to 1.5 V across the entire chip.

Currently, we are using a microcontroller to provide a clock signal to the MEA chip, as well as to control the bias current value. Ideally, this should be performed on-chip to prevent the I/O interface latency, which determines how fast the bias current can be changed and limits the clock's maximum frequency. Figure 10 shows the measured results from preliminary work on the process variation correction PCB system. The output voltage of the chip is digitized and used to control the chip's bias current. At t = 0, since Vout is smaller than 1.5 V, the feedback increases the bias current by 0.1 μA per step, which in turn increases the output DC voltage (Vout ramp at 0.3 s in Figure 10). At 0.32 s, the output DC voltage reaches 1.5 V, and the feedback system locks and holds the bias current constant. This calibration procedure is done once at power-on, and the corresponding Ibias0 values for each input are recorded in the microcontroller for use in the continuous mode testing operation. However, this off-chip method requires an excessive amount of time to adjust the output offset for each individual input during the continuous mode, so the continuous mode feature cannot be operated at the expected 10 kHz rate.

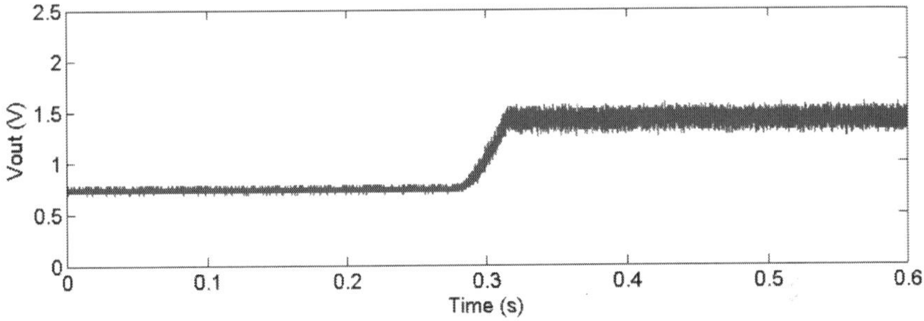

Figure 10. A demonstration of automatic correction for a single channel in stop mode.

4.1.3. Frequency Response and Noise Performance

The frequency response and input-referred noise voltage spectral density of the neural recording IC were measured with both electrical and *in vitro* testing packages using an oscilloscope and a spectrum analyzer, respectively. The stimulus was provided to the IC through an SMA connection. As shown in Figure 11, the system frequency response has a high frequency cutoff of 1.4 MHz. The low frequency corner is below 0.05 Hz, which is beyond the range of our spectrum analyzer, indicating that the pseudo-resistor has a resistance value greater than 0.32 TΩ. The input-referred noise spectrum was measured in stop mode, and the integrated input-referred noise voltage is 4.6 μVrms for a 10 Hz to 10 kHz bandwidth. As a result of the process variation issue, this measurement could not be done in the continuous mode at a 20 kHz sampling rate, due to the length of time required to center the output for each input channel. Therefore, the measurements reported here do not include any potential contribution from noise folding, clock noise, and crosstalk.

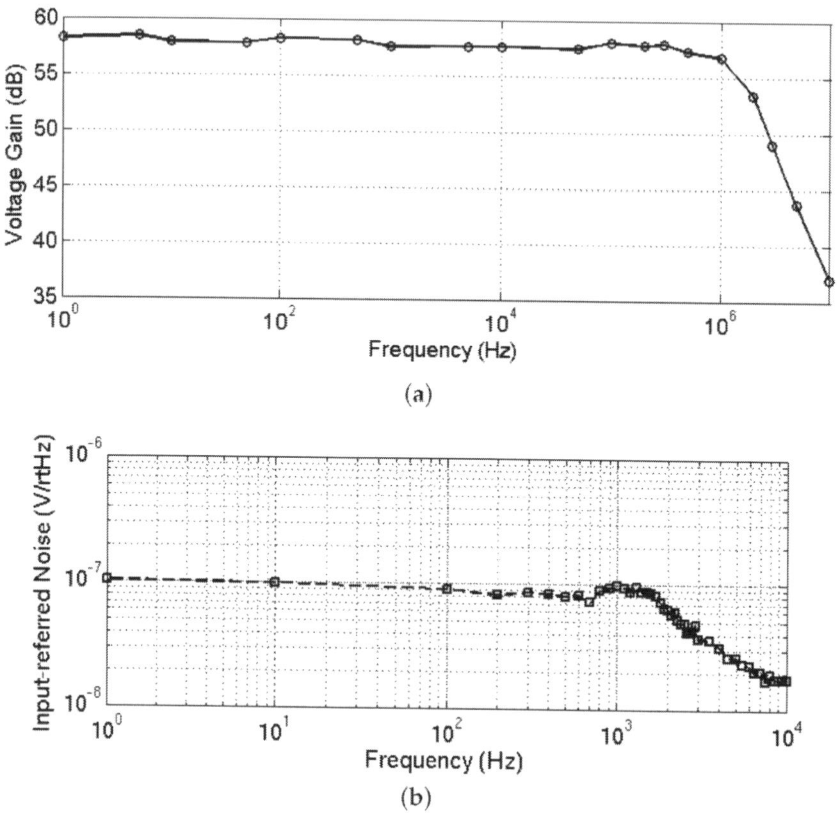

Figure 11. Measured system (**a**) frequency response; and (**b**) input-referred noise.

4.2. Electrical Testing With the In Vitro Package

To record neural signals at the network level, which has significant importance in interpreting brain activities [1], the recording IC has to be sensitive to the signal source locations. To demonstrate the IC's location mapping ability, a 500 Hz sinusoidal signal was applied to a drop of phosphate buffered saline (PBS) solution by a tungsten electrode near the bottom of the chip, with the PBS solution grounded by a silver electrode coated with silver chloride (Ag/AgCl) around the chamber edge. As shown in Figure 12, the amplitudes of the recorded signals decrease as the distance between the C4 recording electrodes and tungsten electrode increases.

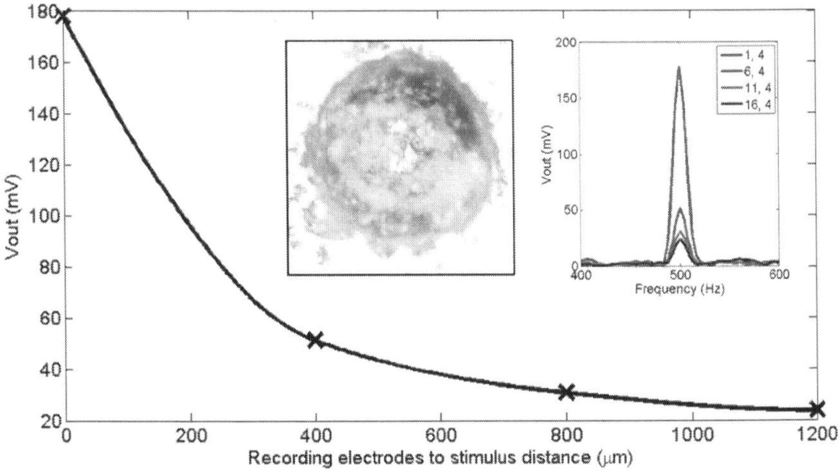

Figure 12. Measured output voltage amplitude with respect to the distance between the recording electrodes and the stimulation electrode. The inset on the right shows the output frequency spectrum, and the inset on the left shows a white substance built up on the surface of a gold plated C4 after electrical testing with the *in vitro* testing package.

4.3. C4 Impedance Measurements

AC impedance measurements were made between the two C4 inputs (1, 3L) and (1, 3R), using the standard 10 × 10 ceramic PGA chip package along with the miniature *in vivo* PCB and flexible cable interface, as shown inFigure 7. The measurements were made with the *in vitro* package by pipetting saline solution within the fluidic chamber on top the C4s to provide conductivity between the two given C4 electrodes. C4 inputs (1, 3L) and (1, 3R) are brought out to external pads and wired via bond wires to I/O pins so that they can be separately accessed. The measurements were made using a low-frequency signal generator to provide a signal input to one C4 and a transimpedance op-amp connected to the other C4 in order to measure the AC current.

Measurements were made on 7–8 multiple-input neuron sensor (MINS) chips over the frequency range of 1 kHz to 10 kHz. Assuming symmetry between both C4s, values of Ri = 725 kΩ and Ci = 620 pF were obtained for the model of a single C4, as shown in Figure 3. The much smaller value of Rs = 3.18 kΩ is seen to have little impact on the overall impedance, given the large value of the interface resistance, Ri. It should be noted that the particular IBM 8HP fabrication run, from which the chips used in this work were obtained, contained leaded C4s. We believe the salt buildup was from incomplete gold-plating and a non-revisable reaction of the solution ions with the leaded C4s. Any future work with a MINS-derivative IC should be done with non-leaded C4s, *i.e.*, Ag/Tin C4s.

5. ACUTE RAT CORTICAL RECORDING

5.1. Surgical Procedure

The fabricated array was tested in a Sprague-Dawley rat (~400 g) under ketamine and xylazine (80 mg/kg and 12 mg/kg, respectively, IP) anesthesia. All procedures were approved and performed in accordance with the guidelines of the Institutional Animal Care and Use Committee, Rutgers University, Newark, NJ. The fur over the skull was shaved, and the animal was placed in a stereotaxic frame. The body temperature was kept at around 36 ∘C with the help of a heating pad under the animal. The blood oxygenation was continuously observed with a pulse oximeter from the hind paw while the animal breathed spontaneously. The skull over the barrel cortex was removed in a 4 mm × 4 mm rectangular area on the right side. The dura was cut and reflected over to expose the cortical surface. Normal saline was applied to keep the cortex moist. The array was attached to a stainless steel rod in the middle of its top surface using medical epoxy (Figure 13). The array was attached to a micromanipulator by the metal rod for easy positioning and mechanical stability. The reference electrode was a platinum wire soldered to the array ground and dipped into the saline pool around the array. The array was slowly lowered until all contacts (C4 bumps) were touching the barrel cortex.

5.2. Recording Procedures

The multiplexed outputs of the recording array were connected to a National Instruments Data Acquisition Board (PCI-6071), and the neural signals were acquired at 20 kHz in 10 s episodes into a desktop computer. Another Matlab code initialized the board and selected one of the 256 contacts for neural recording. Multi-unit activity, as well as LFPs, were recorded as shown in Figure 14a,b, respectively. All the data were collected within an hour after performing the surgical procedure on the rat. Visible C4 imprints were observed on the surface of the rat brain after recording. The control signal (Figure 14d) was recorded with the input grounded through a PCB connector. The control signal is two orders of magnitude smaller than the measured LFP signals.

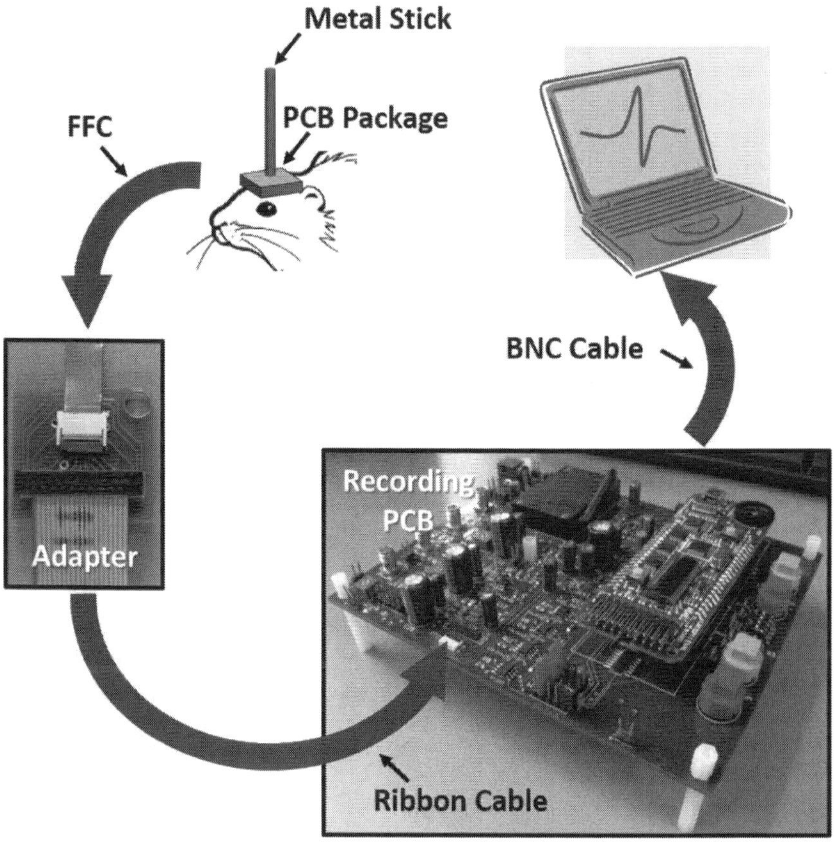

Figure 13. Illustration of the rat cortical recording setup.

6. DISCUSSION AND CONCLUSIONS

An active C4 electrode array for LFP recording has been demonstrated with a mean voltage gain of 58.7 dB and a standard deviation of 0.37 dB after output DC level correction. It has a scalable topology by minimizing the use of on-chip capacitors, while utilizing high-value PMOS pseudo-resistors to extend the low frequency corner. The on-chip C4 3D electrodes are compatible with standard CMOS technology and require minimum post processing. To record low-frequency and low-amplitude neural signals, a sub-Hertz low-frequency corner and 4.6 μVrmsinput-referred noise voltage have been achieved. The power consumption of the chip was minimized to 11.25 mW to prevent neural damage caused by heat. Additionally, the recording chip has been validated with *in vivo*animal recording.

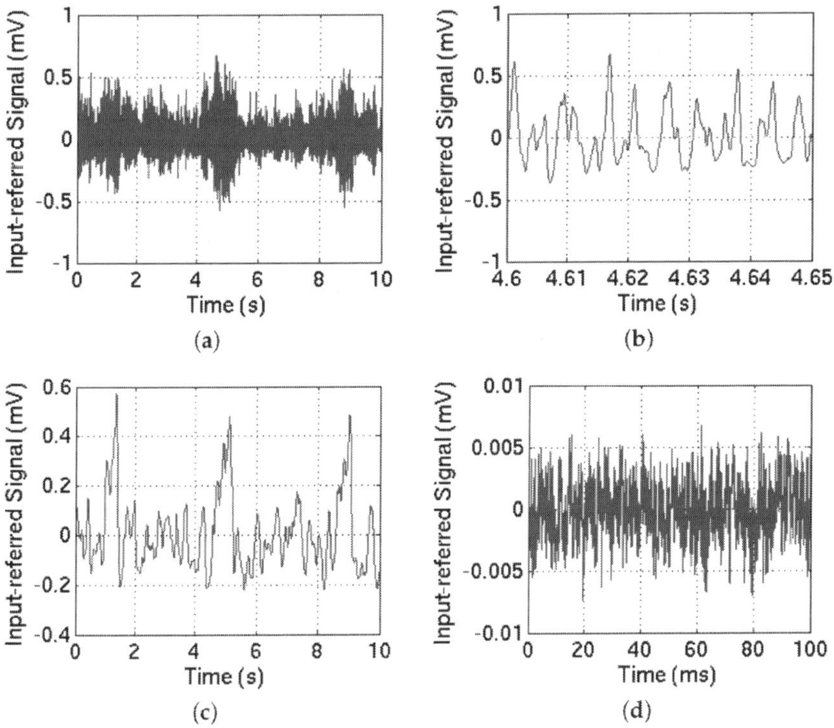

Figure 14. An episode of neural signals recorded in the rat barrel cortex compared to the signal recorded with grounded input. (**a**,**b**): A 300–3000 Hz band-pass filtered version of the signals shows the multi-unit activity—(**b**) is the zoomed view of (**a**); (**c**): A 1–30 Hz band-pass filter was applied to the same data to extract the LFPs; (**d**): The control signal was recorded with the input grounded. A 3000 Hz low-pass filter was applied (The DC voltage was subtracted for clarity of presentation).

In this paper, we demonstrated the use of C4 solder bumps as neural recording electrodes. The MEA was originally designed for both LFP and single unit recording applications. However, our *in vivo* measurements with a live rat were predominately those of LFP signals, apparently due to the distance between the spiking neurons and the C4 electrodes on the surface. For single unit recording use, the chip can be designed with smaller size C4s, when available, or it could be integrated with other penetrating electrodes. Due to the use of a phosphate buffered saline solution during testing , a white substance continually built up on the surface of the gold plated C4s (Figure 12), which significantly increased the resistance values of the C4s when used for a period of time. Eventually, the electrodes became unsuitable for signal recording. So, it is highly recommended to use lead-free C4s, coated with gold, as recording electrodes [36].

Additionally, at this stage, the MINS IC is more suitable for LPF recording applications, since its continuous mode feature can't be operated at the designed 20 kHz sampling rate due to process variation. The future goal is to improve the channel sampling rate in continuous mode, so that the MINS IC can fully take advantage of the spatial resolution of the high density MEA, which is crucial for studying neural function at the network level. Preliminary work (Section 4.1.2) showed that the process variation error can be corrected by adjusting the IC bias current with off-chip real-time data acquisition and feedback circuits. However, the sampling rate is limited by the IC interface, which largely depends on the loading from the IC package and external components. Future iterations of the IC could include on-chip digitally programmable current trimming circuitry to avoid the timing overhead of communicating with off-chip circuits. Thus, it is feasible to scan the entire 256 channels in 50 μs (20 kHz sampling rate) and acquire the spatial information of neural activities recorded in continuous mode.

Finally, it became apparent from our measurements that the ability to adjust the gain of the recording channels would be of considerable value to prevent signals from saturation at the output. Such a gain adjustment could easily be integrated into the design of the MINS IC.

7. FUTURE WORK

For suggested future work, process variation correction circuitry can be integrated on-chip to improve the continuous mode scanning rate and reduce noise coupling to the bias current from the PCB. Also, the gain of the chip could be reduced to 40 dB (or preferably, made adjustable) for a better power/performance tradeoff, to reduce the power density, and to prevent the output signal from saturating for larger input signals. To further reduce the power density, the tail current should be turned off when a column is idle, and turned back on one cycle before the column is selected. Lead-free versions of C4s will be used to replace leaded C4s, and studies on the chronic effects of C4s on living animals will be carried out.

ACKNOWLEDGMENTS

This work was funded by NIH/NINDS under Grant 1R01 NS072385. The chip was fabricated at IBM Microelectronics through a MOSIS Educational Program (MEP)/Research license. The electrical measurements were carried out in Allyn Hubbard's VNNS Lab at Boston University with the help of Howard Cohen. The animal experiment was carried out with Ammar Abdo and Gokhan Ordek's help in Sahin's Neural Interface Lab at New Jersey Institute of Technology. The authors also want to thank Alex Higuera at Boston University for her contribution to the C4 modeling.

AUTHOR CONTRIBUTIONS

Lu Wang, David Freedman, and Ronald Knepper conceived and designed the circuit; M. Selim Ünlü and David Freedman developed the chip post-processing procedure. Lu Wang, David Freedman, and Ronald Knepper designed and performed the electrical experiments; Lu Wang and Mesut Sahin designed and performed the *in vivo* experiments. Lu Wang and David Freedman analyzed the data; Lu Wang, David Freedman, Ronald Knepper, and Mesut Sahin wrote the paper.

REFERENCES

1. Hochberg, L.R.; Bacher, D.; Jarosiewicz, B.; Masse, N.Y.; Simeral, J.D.; Vogel, J.; Haddadin, S.; Liu, J.; Cash, S.S.; van der Smagt, P.; *et al.* Reach and grasp by people with tetraplegia using a neurally controlled robotic arm. *Nature* **2012**, *485*, 372–375.
2. Obien, M.; Deligkaris, K.; Bullmann, T.; Bakkum, D.; Frey, U. Revealing neuronal function through microelectrode array recordings. *Front. Neurosci.* **2015**, *8*.
3. Imfeld, K.; Garenne, A.; Martinoia, S.; Koudelka-Hep, M.; Berdondini, L. Motivations and APS-based solution for high-resolution extracellular recording from in-vitro neuronal networks. In Proceedings of the 3rd International IEEE/EMBS Conference on Neural Engineering, Kohala Coast, HI, USA, 2–5 May 2007; pp. 225–228.

4. Prasad, A.; Sahin, M. Can motor volition be extracted from the spinal cord? *J. NeuroEng. Rehabil.* **2012**, *9*, 372–375.
5. Ordek, G.; Groth, J.D.; Sahin, M. Differential effects of ketamine/xylazine anesthesia on the cerebral and cerebellar cortical activities in the rat. *J. Neurophysi.* **2013**, *109*, 1435–1443.
6. Engel, A.K.; Moll, C.K.E.; Fried, I.; Ojemann, G.A. Invasive recordings from the human brain: Clinical insights and beyond. *Nature Rev. Neurosci.* **2005**, *6*, 35–47.
7. Buzsaki, G.; Anastassiou, C.A.; Koch, C. The origin of extracellular fields and currents—EEG, ECoG, LFP and spikes. *Nature Rev. Neurosci.* **2012**, *13*, 407–420.
8. Shulyzki, R.; Abdelhalim, K.; Bagheri, A.; Salam, M.; Florez, C.; Velazquez, J.; Carlen, P.; Genov, R. 320-Channel Active Probe for High-Resolution Neuromonitoring and Responsive Neurostimulation. *IEEE Trans. Biomed. Circuits Syst.* **2014**, *9*, 34–49.
9. Harrison, R.R. The design of integrated circuits to observe brain activity. *IEEE Proc.* **2008**, *96*, 1213–1216.
10. Eversmann, B.; Jenkner, M.; Hofmann, F.; Paulus, C.; Brederlow, R.; Holzapfl, B.; Fromherz, P.; Merz, M.; Brenner, M.; Schreiter, M.; *et al.* A 128 × 128 CMOS biosensor array for extracellular recording of neural activity. *IEEE J. Solid State Circuits* **2003**, *38*, 2306–2317.
11. Frey, U.; Heer, F.; Pedron, R.; Hafizovic, S.; Greve, F.; Sedivy, J.; Kirstein, K.; Hierlemann, A. An 11k-electrode 126-channel high-density microelectrode array to interact with electrogenic cells. In Proceedings of the IEEE International Solid-State Circuits Conference (ISSCC'07), San Francisco, CA, USA, 11–15 February 2007; pp. 158–159.
12. Imfeld, K.; Neukom, S.; Maccione, A.; Bornat, Y.; Martinoia, S.; Farine, P.A.; Koudelka-Hep, M.; Berdondini, L. Large-scale, high-resolution data acquisition system for extracellular recording of electrophysiological activity. *IEEE Trans. Biomed. Eng.* **2008**, *55*, 2064–2073.
13. Nordhausen, C.T.; Maynard, E.M.; Normann, R.A. Single unit recording capabilities of a 100-microelectrode array. *Brain Res.* **1996**, *726*, 129–140.
14. Frieswijk, T.A.; Bielen, J.A.; Rutten, W.L.C.; Bergveld, P. Development of a solder bump technique for contacting a three-dimensional multi electrode array. *Microsyst. Technol.* **1997**, *3*, 48–52.
15. Aziz, J.N.Y.; Abdelhalim, K.; Shulyzki, R.; Genov, R.; Bardakjian, B.L.; Derchansky, M.; Serletis, D.; Carlen, P.L. 256-channel neural recording

and delta compression microsystem with 3D electrodes. *IEEE J. Solid State Circuits* **2009**, *44*, 995–1005.

16. Song, Y.K.; Borton, D.; Park, S.; Patterson, W.; Bull, C.; Laiwalla, F.; Mislow, J.; Simeral, J.; Donoghue, J.; Nurmikko, A. Active Microelectronic Neurosensor Arrays for Implantable Brain Communication Interfaces. *IEEE Trans. Neural Syst. Rehabil. Eng.* **2009**, *17*, 339–345.
17. Gao, H.; Walker, R.; Nuyujukian, P.; Makinwa, K.; Shenoy, K.; Murmann, B.; Meng, T. HermesE: A 96-Channel Full Data Rate Direct Neural Interface in 0.13 µm CMOS. *IEEE J. Solid State Circuits* **2012**, *47*, 1043–1055.
18. Sodagar, A.M.; Perlin, G.E.; Yao, Y.; Najafi, K.; Wise, K.D. An implantable 64-channel wireless microsystem for single-unit neural recording. *IEEE J. Solid State Circuits* **2009**, *44*, 2591–2604.
19. Bagheri, A.; Gabran, S.; Salam, M.; Perez Velazquez, J.; Mansour, R.; Salama, M.; Genov, R. Massively-Parallel Neuromonitoring and Neurostimulation Rodent Headset With Nanotextured Flexible Microelectrodes. *IEEE Trans. Biomed. Circuits Syst.* **2013**, *7*, 601–609.
20. Gabran, S.; Salam, M.; Dian, J.; El-Hayek, Y.; Velazquez, J.; Genov, R.; Carlen, P.; Salama, M.; Mansour, R. High-Density Intracortical Microelectrode Arrays With Multiple Metallization Layers for Fine-Resolution Neuromonitoring and Neurostimulation. *IEEE Trans. Neural Syst. Rehabil. Eng.* **2013**, *21*, 869–879.
21. Lopez, C.; Andrei, A.; Mitra, S.; Welkenhuysen, M.; Eberle, W.; Bartic, C.; Puers, R.; Yazicioglu, R.; Gielen, G. An Implantable 455-Active-Electrode 52-Channel CMOS Neural Probe. *IEEE J. Solid State Circuits* **2014**,*49*, 248–261.
22. Harrison, R.R.; Charles, C. A low-power low-noise CMOS amplifier for neural recording applications. *IEEE J. Solid State Circuits* **2003**, *38*, 958–965.
23. Yun, X.; Kim, D.; Stanaćević, M.; Mainen, Z. Low-Power High-Resolution 32-channel Neural Recording System. In Proceedings of the 29th Annual International Conference of the IEEE Engineering in Medicine and Biology Society (EMBC'07), Lyon, France, 22–26 August 2007; pp. 2373–2376.
24. Wattanapanitch, W.; Sarpeshkar, R. A Low-Power 32-Channel Digitally Programmable Neural Recording Integrated Circuit. *IEEE Trans. Biomed. Circuits Syst.* **2011**, *5*, 592–602.

25. Han, D.; Zheng, Y.; Rajkumar, R.; Dawe, G.; Je, M. A 0.45 V 100-Channel Neural-Recording IC With Sub-μW/Channel Consumption in 0.18 μm CMOS. In Proceedings of the 2013 IEEE International Solid-State Circuits Conference Digest of Technical Papers (ISSCC), San Francisco, CA, USA, 17–21 February 2013; Volume 7, pp. 735–746.
26. Wright, S.; Polastre, R.; Gan, H.; Buchwalter, L.; Horton, R.; Andry, P.; Sprogis, E.; Patel, C.; Tsang, C.; Knickerbocker, J.; *et al.* Characterization of micro-bump C4 interconnects for Si-carrier SOP applications. In Proceedings of the 56th Electronic Components and Technology Conference, San Diego, CA, USA, 30 May–2 June 2006; pp. 633–640.
27. Wen, S.; Park, K.; Thompson, P.; Shirley, D.; Lee, J.; Park, H. Flip chip assembly with advanced RDL technology. In Proceedings of the 2013 14th International Conference on Electronic Packaging Technology (ICEPT), Dalian, China, 11–14 August 2013; pp. 57–59.
28. Frank, W.; Schenker, I.; Schmutz, P.; Hierlemann, A. Impedance Characterization and Modeling of Electrodes for Biomedical Applications. *IEEE Trans. Biomed. Eng.* **2005**, *52*, 1295–1302.
29. Joye, N.; Schmid, A.; Leblebici, Y. Electrical Modeling of the Cell-Electrode Interface for Recording Neural Activity from High-Density Microelectrode Arrays. *Neurocomput. Elsevier* **2009**, *73*, 250–259.
30. Higuera, A. Neural Interface Model for a Pt/Au C4 Bump Microelectrode Array. Unpublished work. 2016.
31. Wang, L.; Freedman, D.S.; Knepper, R.W.; Ünlü, M.S.; Sahin, M. A 16 × 16 Multi-electrode Array with Integrated CMOS Amplifiers for Neural Signal Recording. In Proceedings of the 2011 IEEE Biomedical Circuits and Systems Conference (BioCAS), San Diego, CA, USA, 10–12 November 2011; pp. 82–85.
32. Carusone, T.C.; Johns, D.A.; Martin, K.W. *Analog Integrated Circuit Design*, 2nd ed.; John Wiley & Sons: Hoboken, NJ, USA, 2011.
33. Majidzadeh, V.; Schmid, A.; Leblebici, Y. A 16-channel 220 μW neural recording IC with embedded delta compression. In Proceedings of the 2011 IEEE Biomedical Circuits and Systems Conference (BioCAS), San Diego, CA, USA, 10–12 November 2011; pp. 9–12.
34. Hosseini-Nejad, H.; Jannesari, A.; Sodagar, A. Data Compression in Brain-Machine/Computer Interfaces Based on the Walsh-Hadamard Transform. *IEEE Trans. Biomed. Circuits Syst.* **2014**, *8*, 129–137.
35. Yazdani, N.; Rashidi, A.; Sodagar, A.M.; Mohebbi, M. Reduction of spatial data redundancy in implantable multi-channel neural recording

microsystems. In Proceedings of the 2014 IEEE Biomedical Circuits and Systems Conference (BioCAS), Lausanne, Switzerland, 22–24 October 2014; pp. 208–211.

36. Berdondini, L.; Wal, P.D.V.D.; Rooij, N.F.D.; Koudelka-Hep, M. Development of an electroless post-processing technique for depositing gold as electrode material on CMOS devices. *Sens. Actuators B Chem.* **2004**, *99*, 505–510.

37. Aziz, J.N.Y.; Genov, R.; Bardakjian, B.L.; Derchansky, M.; Carlen, P.L. Brain-silicon interface for high-resolution *in vitro* neural recording. *IEEE Trans. Biomed. Circuits Syst.* **2007**, *1*, 56–62.

38. Graham, A.H.D.; Bowen, C.R.; Surguy, S.M.; Robbins, J.; Taylor, J. New prototype assembly methods for biosensor integrated circuits. *Med. Eng. Phys.* **2011**, *33*, 973–979.

Chapter 6

Design and Implementation of a RF Powering Circuit for RFID Tags or Other Batteryless Embedded Devices

Dongsheng Liu, Rencai Wang*, Ke Yao, Xuecheng Zou and Liang Guo

School of Optical and Electronic Information, Huazhong University of Science & Technology, Wuhan 430074, China

ABSTRACT

A RF powering circuit used in radio-frequency identification (RFID) tags and other batteryless embedded devices is presented in this paper. The RF powering circuit harvests energy from electromagnetic waves and converts the RF energy to a stable voltage source. Analysis of a NMOS gate-cross connected bridge rectifier is conducted to demonstrate relationship between device sizes and power conversion efficiency (PCE) of the rectifier. A rectifier with 38.54% PCE under normal working conditions is designed. Moreover, a stable voltage regulator with a temperature and voltage optimizing strategy including adoption of a combination resistor is developed, which is able to accommodate a large input range of 4 V to 12 V and be immune to temperature variations. Latch-up prevention and noise isolation methods in layout design are also presented. Designed with the HJTC 0.25 μm process, this regulator achieves 0.04 mV/°C temperature rejection ratio (TRR) and 2.5 mV/V voltage rejection ratio (VRR). The RF powering circuit is also fabricated in the HJTC 0.25 μm process. The area of the RF powering circuit is 0.23×0.24 mm^2. The RF powering circuit is successfully integrated with ISO/IEC 15693-compatible and ISO/IEC 14443-compatible RFID tag chips.

Keywords: batteryless embedded devices; wireless sensor network; RFID tag chip; rectifier; regulator

1. INTRODUCTION

The applications of RFID technologies, including supply chain management, access control to buildings, public transportation, airport baggage handling, express parcel delivery and logistics, have increased rapidly in recent years [1–3]. Owing to its wireless and far field measurement capabilities, the combination of RFID technology and wireless sensing networks has also found a large number of applications such as environmental monitoring and healthcare devices [4–6]. RFID tags are often classified as passive or active. The passive tag is powered by an electromagnetic RF (radio frequency) wave transmitted by a reader, while the active tag is powered by a battery. The passive tag has advantages of low cost and long life, thus enable sensing nodes in a sensing system to be placed in positions that are not easily accessible [7,8].

However, passive RFID tags, and other batteryless embedded devices such as contactless smart cards, wireless sensor nodes and implantable biomedical devices, have to work at a considerably long distance from the transmitter [9,10]. Moreover, power converted from the electromagnetic RF wave to the tag should be sufficient to supply device in its operating distance, especially when a power hungry sensor is integrated. As the RF energy received by the tag (transponder) or other batteryless embedded devices decreases rapidly with distance, the induced voltage across the tag antenna is often variable in big interval and sensitive to temperature and process variation [11,12].

Thus, key challenges to implement a passive tag are harvesting energy with high efficiency and generating a stable voltage source for a digital baseband controller, the memory of the tag and its functional parts. This voltage source must be immune to process, input voltage and environmental temperature changes. Meanwhile, in order to meet the requirements for a more complicated RFID tag such as integrating sensors and CPU in a tag chips, the RF power circuit should be capable of handling heavy current loads [13,14].

To solve those problems, many RF power circuit design techniques for passive RFID tags have been reported to improve the power conversion efficiency and stability of the regulator power supply [14–17].

A low power RF harvester for a smart passive sensor tag used in structural health monitoring applications was proposed in [16]. The RF harvester harvests energy from the RF field and provides a stable and controllable power supply. Measurements of this design indicate full operation of the sensor tag for time intervals of 20 ms for up to 4 m distances with a tag reader power of 200 mW. However, this RF harvester is a chipless design while tags that utilize an application-specific chip are an undeniable trend for passive tag applications.

For specific tag chip designs, a low power CMOS voltage regulator for a wireless blood pressure biosensor is presented in [8]. A low voltage regulator, as part of the associated electronics of a blood pressure bio-medical implanted device, provides the power supply for the system. The line regulation of this regulator is 39 mV/V and its total power consumption is 1.2 mW. Since the body temperature is kept stable at approximately 37 °C, the regulator of this design is not optimized for temperature changes.

Moreover, to improve the stability of the power supply, a RF powering circuit which is immune to variations of temperature and voltage is proposed in [14]. A ripple stabilizer and a temperature stabilizer have been integrated to maintain a stable reference voltage that is insensitive to temperature variation and voltage ripples. Its line regulation is 12 mV/V and temperature rejection ratio is 0.2 mV/°C. However, the two huge resistors used for meeting the requirement of the low static current occupy a large area.

As for achieving high PCE, a RF to DC CMOS rectifier with high efficiency over a wide input power range for RFID applications is implemented in [17]. The rectifier is based on a novel active load circuit which adjusts the output current as a function of the incoming RF power. This design allows maximizing both the efficiency and sensitivity of the circuit and achieves a highest PCE of 45%.

In this paper, a RF powering circuit which is immune to process, voltage and temperature changes with high PCE is designed for a HF passive RFID tag chip. Compared with previous works, the input voltage of the regulator ranges from 4 V to 12 V which is much larger than that reported in the previous works, while a stable output voltage is also maintained against temperature and voltage variations. Furthermore, a rather high PCE is achieved in this work, which means this RF powering circuit can provide more power to the tag and hence enable further identification distance and it can be applied in more complicated passive tags.

The paper is organized as follows: Sections 2 and 3 describe the design and implementation of the RF powering circuit. The simulation and measurement results are presented in Section 4 and followed by conclusions in Section 5.

2. ARCHITECTURE OF THE RF POWERING CIRCUIT

A passive RFID tag generally consists of an analog front end (AFE), a digital baseband and an EEPROM, where the EEPROM can be divided into a logic control part and a memory part. If the tag is integrated with other specific functions such as sensing or monitoring, a functional block can be added to the tag system. In this design, the digital baseband and the logic control part of EEPROM are realized with standard cells, while the analog design is responsible for the AFE and the memory part of EEPROM.

As shown in Figure 1, the RF powering circuit comprises four parts. First, the energy harvesting circuit, consisting of an inductor and a resonance capacitor, harvests energy from an electromagnetic field through inductive coupling and resonance. A magnetic coil is used as antenna in this design. Then, the rectifier circuit, which consists of a bridge rectifier and a capacitor, converts the input RF signal from electromagnetic field into a DC signal with ripples. Consisting of a bias circuit and regulator V_{CC}/V_{DD},the regulator is designed to convert the DC signal with ripples into a stable power supply. The output voltage of the regulator has much smaller ripples and can drive several loads. Finally, an over-voltage protection circuit is presented to limit voltage of antennas by bleeding spare energy in antennas into the ground and this ensures that the whole chip has a stable and reliable operation.

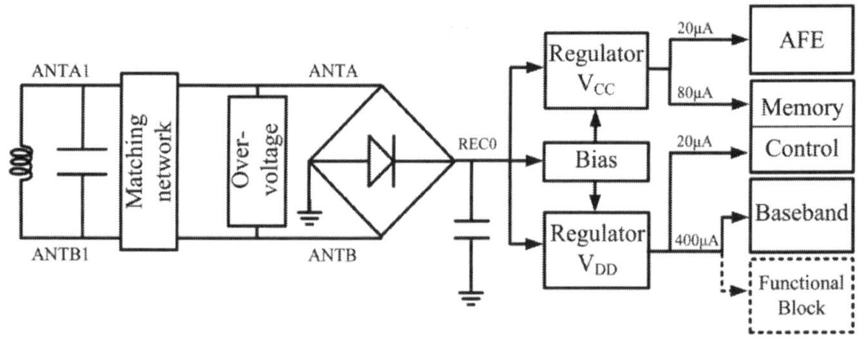

Figure 1. Structure of the RF powering circuit.

The regulator V_{CC} drives the AFE and the memory part of the EEPROM. The average load current of the regulator V_{CC} is 100 μA, in which 80 μA is for the memory array and 20 μA is for the AFE. The regulator V_{DD}drives the digital baseband controller and the logic control of the EEPROM. The load current of the regulator $V_{DD}$is 420 μA, where 20 μA is for the EEPROM logic control, and 400 μA is for baseband controller and other functional blocks.

In our design, the digital and analog circuits are driven by V_{DD} and V_{CC} respectively, which makes it flexible to design the driving capability for different circuits, therefore meeting the requirements of high power consumption chips such as digital baseband controllers based on FSM, security chips, CPUs and chips integrated with sensors. Meanwhile, the huge noise of the digital circuit can be isolated by driving the digital and analog part separately. Thus the cost of noise restriction in the AFE can be minimized in this way.

3. DESIGN AND IMPLEMENTATION OF RF POWERING CIRCUIT

3.1. Rectifier Circuit

As shown in Figure 2, a NMOS gate cross-connected bridge rectifier structure, which is usually used in RF transponders for its high PCE, is adopted as the rectifier circuit.

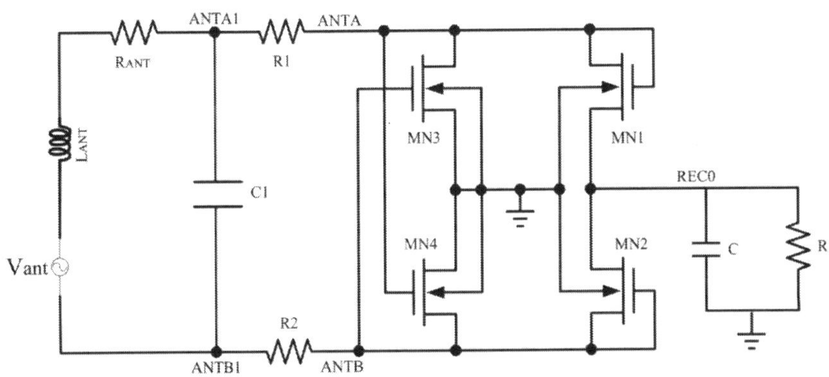

Figure 2. NMOS gate cross-connected bridge rectifier.

In Figure 2, the gates of NMOS transistors MN3, MN4 are cross connected to the antennas and MN1, MN2 are connected as diodes. The antenna can be equivalent to an AC voltage source V_{ant}, an inductor L_{ANT} and the parasitic resistor R_{ANT}. Resistors R1, R2 form the matching circuit, which matches the impedances of the antenna and the rectifier, thus L_{ANT} and C1 resonate and the maximum voltage values of ANTA and ANTB will be obtained. The impedance of R1 and R2 is obtained according to the input impedance of the cascading circuit to the resistors R1 and R2.

The equivalent circuit of the NMOS gate cross-connected bridge rectifier is shown in Figure 3a. Resistors R1, R2 form the matching circuit, and R_L is the equivalent input impedance of the cascading circuit to the resistors R1 and R2, V_{ANT} is the input voltage of the rectifier. Resistors R1, R2, and R_L are connected to C1 in parallel. They can be equivalently transformed to be connected in series as shown in Figure 3b. In Figure 3b, R_{LS} is obtained after the parallel to serial impedance conversion of resistors R1, R2, and R_L. An approximative value of R_{LS} is calculated by the equation below:

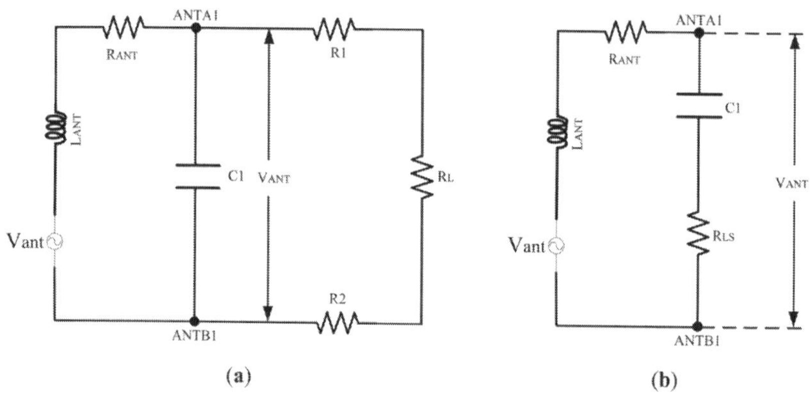

Figure 3. (**a**) Equivalent circuit of the NMOS gate cross-connected bridge rectifier; (**b**) Equivalent circuit of (a) after parallel to serial impedance conversion.

$$R_{LS} = \frac{1}{(R_L + R_1 + R_2)\cdot(\omega C1)^2} \quad (1)$$

According to the series resonance theory, the quality factor Q of the circuit in Figure 3b can be obtained by:

$$Q = \frac{1/\omega C1}{1/(R_1 + R_2 + R_L)\cdot(\omega C1)^2 + R_{ANT}} \quad (2)$$

When the value of Q is greater, the input voltage V_{ANT} will be higher. According to Equation (2), to make Qbigger, the values of R1 and R2 should be big as well. As the value of R_{ANT} is 20 Ω, the value of C1 is 16.5 pF, so the values of resistors

R1 and R2 are both designed to be 15.7 kΩ to make the previous factor of the denominator of Equation (2) infinitesimally smaller than the following one.

When ANTA is high and ANTB is low, MN1, MN4 are on and MN2, MN3 are off. The capacitor is charged through path ANTA-MN1-C-GND-MN4-ANTB. When ANTB is high and ANTA is low, the capacitor will be charged through path ANTB-MN2-C-GND-MN3-ANTA.

In this circuit, the threshold voltage loss is as low as a V_{TH}, which is the threshold voltage of NMOS. As MN3 and MN4 do not produce any V_{TH} loss, the threshold voltage loss is produced by MN1 and MN2. It can be calculated that the minimum operation input level is slightly higher than V_{TH} because of the substrate effect. Thus, the minimum operation input level is efficiently reduced in this design. According to the preceding analysis, the antenna voltage is similar to a sinusoidal signal, and the output voltage level is [18]:

$$V_{RECO} = V_{ANT} / \sqrt{2} - V_{TH} \tag{3}$$

where V_{RECO} is the output voltage of the rectifier, V_{ANT} is the peak amplitude of the antennas voltage, V_{TH} is the threshold voltage of NMOS. As $V_{SB} = V_{RECO}$, V_{SB} is the source-bulk voltage of MN1 and MN2, according to the bulk effect model of NMOS, there is:

$$V_{TH} = V_{TH0} + \gamma(\sqrt{|2\phi_F + V_{RECO}|} - \sqrt{2\phi_F}) \tag{4}$$

where V_{TH0} is the intrinsic threshold voltage, γ is the body effect coefficient, ϕ_F is the Fermi potential. According to Equations (3) and (4), V_{RECO} increases with respect to V_{ANT}. As the electromagnetic field strength changes from 150 mA/m to 5 A/m, which is defined in ISO/IEC 15693 standard and related ISO/IEC 10373-7 test standard, the output voltage of the rectifier varies a lot from 4 V to 12 V according to the simulation. As a rectifier converts a RF signal of antenna to a DC voltage, higher PCE of the rectifier means more energy is converted to the tag. PCE is calculated by the equation below:

$$PCE = \frac{P_{out}}{P_{in}} \tag{5}$$

PCE can be also calculated by $PCE = P_{out}/(P_{out} + P_{mos})$, where P_{mos} is the power consumption of the transistors of the rectifier. P_{mos} can be calculated by, $P_{mos} = P_{dynamic} + P_{static}$, where $P_{dynamic}$ is the power consumption of the transistors while the transistors are on, and P_{static} is the power consumption of the transistors while they are off. $P_{dynamic} = \overline{V_{on} \cdot I_{on}}$ V_{on} is the forward voltage drop of the transistors, and I_{on} is the forward current of the transistors $P_{static} = \overline{V_{reverse} \cdot I_{reverse}}$, $V_{reverse}$ is the reverse voltage of the transistors, and $I_{reverse}$ is the reverse leakage current of the transistors. The turn-on voltage of the rectifier determines the minimum input-voltage of the rectifier, higher turn-on voltage means higher Von, and greater reverse leakage current means greater P_{static}, so while the turn-on voltage is higher, and the reverse leakage current is greater, then the input power will be greater.

According to the aforementioned analysis of resistors R1 and R2, the resistors make PCE higher. As proper R1 and R2 bring a biggest Q of the circuit, so the value of V_{ANT} will be maximum when L_{ANT} and C1 resonate. Bigger V_{ANT} means greater P_{out}, then the value of PCE will be higher.

As shown in Table 1, PCE is greatly influenced by the size of the MOS device. According to the ISO/IEC 15693 standard and the related ISO/IEC 10373-7 test standard, the input voltage of the rectifier is always higher than 4 V, so the V_{ANT} is set to be 6 V as a normal value in this simulation. The load current of this RF powering circuit is less than 600 μA, and the output power of the antenna is normally less than 5 mW, so the R_{load} is set to be 6 kΩ. In this table, MOS devices have the same width/length (represented as W/L afterwards) ratio and M is the multiplier of a MOSFET. Greater M value means bigger size of a MOSFET. Bigger size of a rectifier MOSFET brings smaller channel conduction resistor, thus the output voltage of the rectifier is increased and higher PCE is achieved. However, bigger size of a MOSFET increases Ids of the MOS device, hence it consumes more power. As a result, when $M > 20$, the increase of PCE becomes slower than that when M is smaller. Tradeoffs between PCE and power consumption should be considered when designing the rectifier.

Table 1. Simulation results of NMOS gate cross-connected bridge rectifier (R_{load} = 6 kΩ, V_{ANT} = 6 V).

M	VHD/V	Pin/mW	Pout/mW	PCE
5	2.64	2.60	1.16	44.60%
10	3.01	3.00	1.51	50.23%
20	3.30	3.30	1.83	55.30%
30	3.46	3.53	2.00	56.60%
40	3.55	3.68	2.10	57.05%
50	3.62	3.78	2.18	57.70%

3.2. Linear Regulator

The V_{CC} voltage regulator is shown in Figure 4. By using a combination resistor and a calculation strategy for the device parameters, an optimization for regulator performance related to voltage and temperature is achieved in this work. As can be seen in Figure 4, NMOS transistors MN1 and MN2 form a current mirror. The size of MP1 is two times that of MP2. The gates of MP5 and MN3 are connected. Capacitance C1 and resistor R0 are frequency compensation devices. MP0 is a large pass transistor which enables heavier load capacity. MP4, MP5 and R2 stabilize the V_{CC} voltage level while C0 is used for filtering and storing energy. The voltage limiter of this circuit is composed of MP6, MP7, MN5, R1 and MN4.

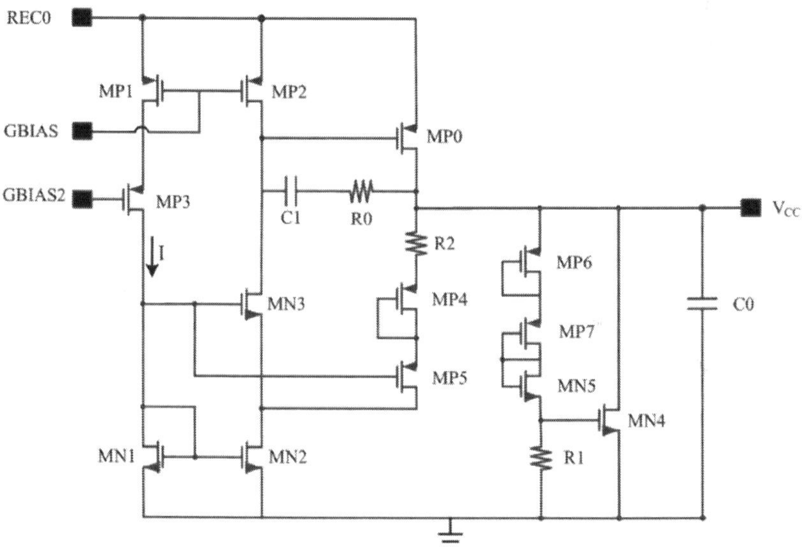

Figure 4. Circuit of V_{CC} regulator.

If calculating the output voltage through the way R2-MP4-MP5-MN1, the output voltage of the regulator inFigure 4 is given by [13]:

$$V_{CC} = V_{SG,MP4} + V_{SG,MP5} + I_{D,MP5} R_2 + V_{SG,MN1} \tag{6}$$

where MP4, MP5 and MN1 are in saturation region with the same *W/L* ratio, according to the V-I characteristic of MOSFET in saturation region, the following equation can be given as:

$$V_{SG,MP5} = V_{SG,MP4} = |V_{THP}| + \sqrt{\frac{2I_{D,MP5}}{\mu_P C_{ox}(W/L)_{MN1}}} \tag{7}$$

$$V_{SG,MN1} = V_{THN} + \sqrt{\frac{2I_{D,MN1}}{\mu_N C_{ox}(W/L)_{MN1}}} \tag{8}$$

V_{THN} and V_{THP} are the threshold voltages of NMOS and PMOS respectively.

As the size of MP1 is two times that of MP2, MN1 and MN2 are mirror transistors, and the current can be represented as:

$$2I_{D,MP5} = 2I_{D,MP2} = I_{D,MP1} = I_{D,MN1} \tag{9}$$

Equation (6) can be rewritten as:

$$V_{CC} = |V_{THN}| + 2|V_{THP}| + 2\sqrt{\frac{2I_{D,MP1}}{\mu_P C_{ox}(W/L)_{MN1}}} + \sqrt{\frac{2I_{D,MP1}}{\mu_N C_{ox}(W/L)_{MN1}}} + \frac{1}{2}I_{D,MP1}R_2 \tag{10}$$

In Equation (8) $I_{D,MP1}$ is determined by the bias circuit with the value of 2 µA. When ignoring the changes of temperature, the first and second term at the right side of Equation (10) are constant, whereas the last three terms are the only part that can be changed in this equation. Suppose that the last three terms of Equation (10) are equal. That is:

$$2\sqrt{\frac{2I_{D,MP1}}{\mu_P C_{ox}(W/L)_{MN1}}} = \sqrt{\frac{2I_{D,MP1}}{\mu_N C_{ox}(W/L)_{MN1}}} = \frac{1}{2}I_{D,MP1}R_2 = \frac{1}{3}(V_{CC} - |V_{THN}| - 2|V_{THP}|) \tag{11}$$

According to Equation (11), the value of $(W/L)_{MP1}$ and R_2 can be calculated under typical process conditions. The output voltage of the regulator changes along with the environment temperature and the rectified voltage of RECO. When the voltage of RECO increases, $V_{SG,MP1}$ and $V_{SG,MP2}$ are clamped by the bias circuit. Since MP3 is a common-gate connection PMOS, the voltage change of RECO mostly turns out to be the drain-source voltage of MP3, so the variation of drain voltage of MP1 and the current mirror mismatch of MP1 and MP2 caused by channel modulation effect are reduced. Similarly, the matching of MN1 and MN2 is improved by MN3. Thus, the immunity of the regulator output to voltage changes of RECO are improved, and according to the analysis of the bias circuit mentioned later in Section 3.3, the current I of this regulator is not easily affected by the supply voltage of RECO, so fluctuations of V_{CC} caused by the supply voltage can be suppressed by a well-designed bias circuit.

Temperature Coefficient (TC) is an index which describes how temperature variations influence a specific value. It is mostly used to describe changes of resistance values. TC of this design is defined as Equation (12) $V_1$and V_2 are voltages before and after the temperature change:

$$TC = \frac{V_2 - V_1}{V_1 \cdot \Delta T} \tag{12}$$

By observing the right side of Equation (10)V_{THN}, V_{THP}, $\mu_P C_{ox}$ and $\mu_N C_{ox}$ are process parameters and their contributions to V_{CC} are fixed when the *W/L* ratio is determined. TC of the fifth item $1/2 I_{D,MP1} R_2$ can be modulated by selecting the proper type of resistor. TC of the front four items of Equation (10) around room temperature can be obtained by simulation. According to the obtained TC, a proper type of resistor for R2 can be selected. In this way, TC of V_{CC} is minimized around room temperature.

While types of resistors are limited in each process, the TC of resistors which can be chosen is also limited. To solve this problem, a concept of combination resistor is adopted in our design [19]. According to this concept, two resistors of different TC are chosen to compose a combination resistor. TC of V_{CC} can be minimized by choosing a proper TC of this combination resistor. As the TC of the combination resistor is between that of the chosen resistors, the combination resistor has many more TC options than a single resistor. In general, by setting a proper voltage proportion of the first three items in Equation (8) and selecting a combination resistor appropriately, a desirable V_{CC} voltage regulator circuit immune to temperature variations is designed in this work. Moreover, the circuit of the V_{DD} regulator has the same structure with that of V_{CC}, while loads of the output voltage are the main differences between the two, one can carry on the design and optimization of V_{DD}regulator with the same methods as V_{CC} mentioned above.

3.3. Bias Circuit

Shown in Figure 4, a bias circuit is designed to provide a stable current for the regulator and other circuits. Start-up part of this bias circuit is composed of PMOS transistors MP1 and MP2. In the process of starting up, if there are no transistors MP1 and MP2, the bias circuit will enter a condition that the voltage of GBIAS1 is high and the gate voltage of MN1 and MN2 is low when RECO goes up. In this condition, there is no current in each branch of the bias circuit and the bias circuit isn't able to start to work. When starting up, the rectified voltage RECO goes up and the voltage of GBIAS1 rises. At first, transistors MP1 and MP2 are both off and capacitor C1 is uncharged. Then MP2 turns on while RECO goes high, thus making the voltage of GBIAS2 increase. Then transistors MN1 and MN2 turn on, reducing the drain voltage of MN2, thus making transistors MP5 and MP1 on. Capacitor C1 begins to charge through MP1. Hereafter, the bias circuit enters its normal working conditions when the upper plate of C1 is charged to RECO. Then MP2 turns off and the start-up circuit is out of operation afterwards.

As for the bias circuit itself, the current mirror in Figure 5 is composed of PMOS transistors MP4 and MP5. The*W/L* ratio of NMOS transistor MN2 is k times of MN1. MOS devices in Figure 5 are all in saturation region. As MP4 and MP5 are mirror transistors of each other, the current of each branch (represented as I_{bias}) is kept equal. There has the following equation [20]:

$$V_{GS,MN2} + I \cdot R = V_{SG,MP1} \tag{13}$$

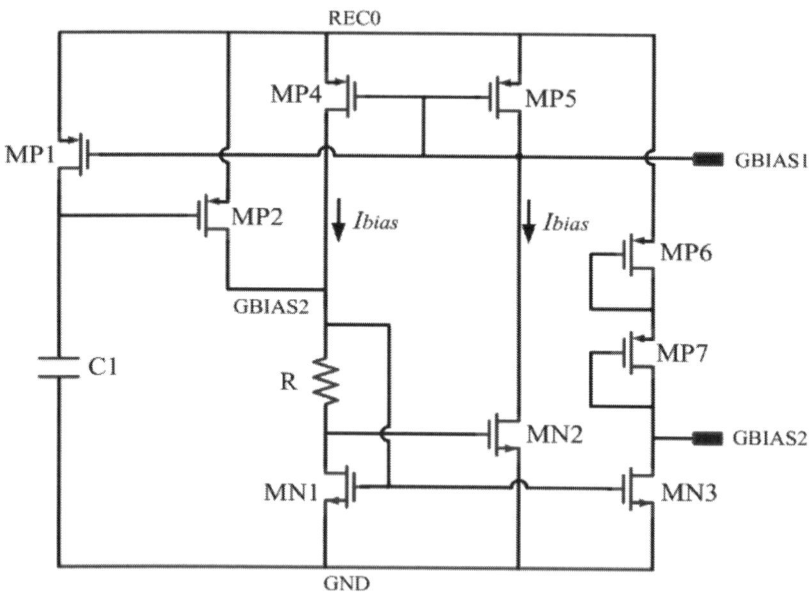

Figure 5. Bias circuit of RF powering circuit.

Since it is known that MN3 and MN4 are in saturation region, according to the V-I characteristic of MOSFET in saturation region, I_{bias} can be calculated from Equation (13):

$$I_{bias} = \frac{2}{\mu_N C_{ox} (W/L)_{MN1} \cdot R_1^2} (1 - \frac{1}{\sqrt{k}})^2 \tag{14}$$

Shown in Equation (14), the advantage of this bias circuit is that it can eliminate the substrate bias effect. The *W/L* ratio of MN1, MN2 and the resistance value of *R* can be calculated according to the requirements of power dissipation, as well as the proper selection of over-drive voltage of MN1 and MN2.

As long channel devices are used in the bias circuit, the channel length modulation effect produced by the increasing of V_{RECO} can be well suppressed. Moreover, considering the temperature response of this circuit, Equation (15) can be obtained when taking the derivative of temperature in Equation (14):

$$\frac{\partial I}{\partial T} = A(\frac{1}{\mu_N} \cdot \frac{\partial \mu_N}{\partial T} + \frac{2}{R} \cdot \frac{\partial R}{\partial T}) \tag{15}$$

While $\partial \mu_N / \partial T$ is a fixed process coefficient, *A* is a fixed coefficient, and *R* represents the impedance of the resistor in Figure 5. By choosing proper value of $\partial R / \partial T$, the TC of bias current, expressed in Equation (15), can be minimized around room temperature. The combination resistor mentioned above is also adopted here.

3.4. Layout Optimization Methods

In this work, the RF powering circuit is designed for application in a passive RFID tag. As the tag is integrated with digital and analog parts in the same chip, the noise interference between the digital and analog parts becomes a serious problem as it operates under high frequency. As the RFID tag can work in intense electromagnetic field according to the related ISO/IEC 15693 and ISO/IEC 14443 standards, large spare energy in antennas must be bled by an over-voltage protection circuit, thus a large current is injected into the ground. At the same time, the high frequency of transient currents is conducted in the rectifier circuit.

These problems listed above can be latch-up triggers in both the analog and digital part, therefore leading to a latch-up in the chip which is a dominant failure in CMOS chip projects. Latch-up may cause a loss of data logic states or destructive failure of the whole system [21,22]. As a result, layout optimizations to isolate the noise between the analog and digital part and to reduce the possibility of failures caused by latch-up is quite essential in the layout design of this chip.

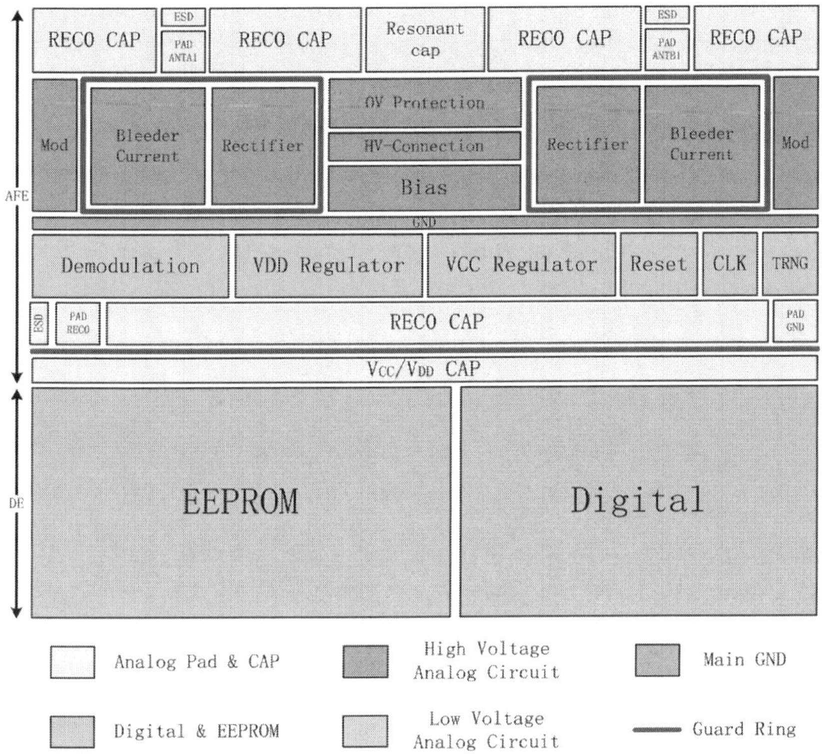

Figure 6. Layout floor plan of the whole chip.

The plan of the whole chip layout is shown in Figure 6. The AFE and DE part of the chip are separated with V_{RECO} and V_{CC}/V_{DD} capacitances so as to isolate noise between them. Since RECO is designed to be 4 V–12 V, which is much higher than the operation voltage of DE, an NP guard ring is placed between RECO and V_{CC}/V_{DD}capacitances to split the high voltage AFE and low voltage DE areas. Moreover, in the AFE layout, high voltage blocks are apart from low voltage blocks with a main thick ground wire to avoid latch-up. To isolate substrate current in areas with large current flow, an NPN guard ring is added around the current bleeding and rectifier circuits for large currents injected into the substrate that increase the ground voltage which is more likely to lead to a latch-up.

For more considerations, two antenna pads ANTA1 and ANTB1 are placed as mirror images in this chip layout shown Figure 6, as well as wires, devices and circuits connected to the two antennas. Signals transferred through the two antennas are ensured to be symmetrical in this way. Buffer cells are added into the long wires connecting the AFE and DE circuits to suppress signal attenuations.

4. SIMULATION AND MEASUREMENT RESULTS

The RF powering circuit is designed and simulated in the HJTC 0.25 μm process together with the whole RFID tag. As mentioned in Section 3, PCE of the RF powering circuit designed in HJTC 0.25 μm process is simulated under different load current of V_{CC}/V_{DD}. Simulation results are shown in Table 2, PCE of the circuit is 39.17% when the load current is 520 μA while the tag is in its normal working conditions.

Table 2. Simulation results of PCE of RF powering circuit in HJTC 0.25 μm process.

V_{RECO}	I_{load}	$V_{ant,p\text{-}p}$	PCE
4.42 V	800 μA	7.03 V	53.14%
4.88 V	600 μA	7.36 V	43.68%
5.01 V	520 μA	7.43 V	39.17%
5.2 V	400 μA	7.53 V	31.94%

Temperature rejection ratio (TRR) and voltage rejection ratio (VRR) are defined as follows:

$$TRR = \frac{\Delta V_{CC}}{Temper} \quad (16)$$

$$VRR = \frac{\Delta V_{CC}}{V_{olt}} \quad (17)$$

where ΔV_{CC} is the variation of regulator output, Temper is the variation range of temperature; V_{olt} is the variation range of rectifier output voltage.

Shown in Figure 7 are the simulation results of regulator V_{CC} in the HJTC 0.25 μm process. The calculations of the simulation results are shown in Table 2. The value of PCE can be calculated with the equation $PCE = P_{out}/P_{in}$, while $P_{in} = \overline{V_{ant} \cdot I_{ant}}$ and $P_{out} = \overline{V_{CC} \cdot I_{load}}$, is the voltage of the antenna, I_{ant} is the current of the antenna, V_{CC} is the output voltage of the regulator, I_{load} is the load current of the regulator. Considering the usual work environment of the tag, we pay more attention to the value of TRR while V_{RECO} is near 6 V, and the value of VRR while the temperature is near 27 °C. Under these two conditions the regulator has a quite small TRR and VRR with TRR = 0.04 mV/°C and VRR = 1.1 mV/V.

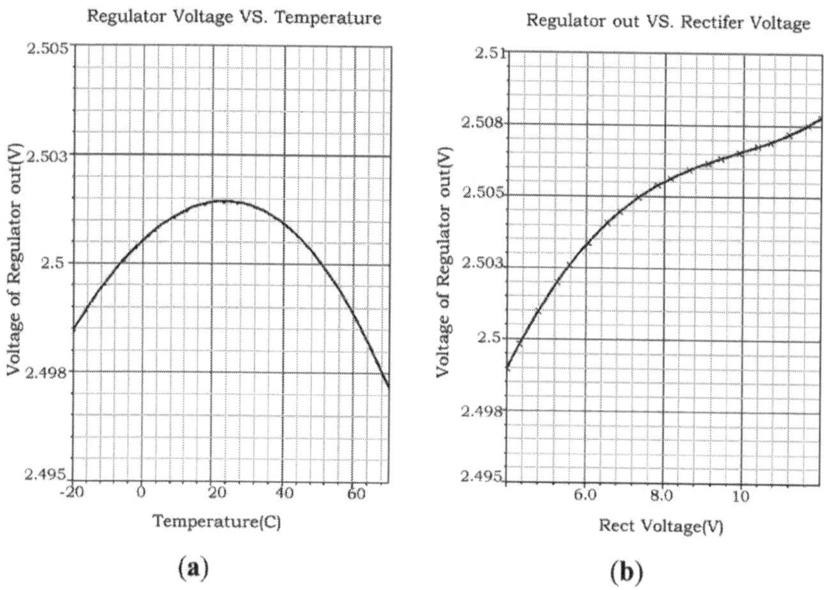

Figure 7. (**a**) V_{CC} *vs.* Temp Simulation results of regulator V_{CC}, V_{RECO} = 6V; (**b**) V_{CC} *vs.* V_{RECO} Simulation results of regulator V_{CC}, T = 27 °C.

RFID tags designed in this paper are also measured according to ISO.IEC 15693 standard and related ISO/IEC 10373-7 test standard with contactless test methods. The test electromagnetic field strength (represented as H) defined in the standard is 150 mA/m (H_{min} = 150 mA/m) to 5 A/m (H_{max} = 5 A/m), and test simulation signals are in 100% ASK or 10% ASK modulation mode.

The RF powering circuit has been applied to HF RFID tag chip projects. To make the testing more convenient, the analog front end (AFE) circuit for a HF RFID tag is fabricated, in which this RF powering circuit is integrated. Consisting of analog front end (AFE), digital baseband and EEPROM memory (DE), the RFID tag test platform based on PCB is shown in the Figure 8. The DE part is realized in the form of FPGA, the AFE part is fabricated in the HJTC 0.25 μm process. The RF powering circuit designed above is one part of the AFE in this chip. The partial enlarged view in Figure 8 shows that the distance from reader to the tag can be adjusted to test different electromagnetic field strengths. In Figure 8, the demodulated signal coming from the AFE chip and the return signal which is transformed from the demodulated signal just mentioned via the processing of the DE part are tested by the oscilloscope. In the oscilloscope screen shown in Figure 8, the upper curve is the test waveform of the demodulated signal, while the lower one is the test waveform of the return signal. Both of the test waveforms are normal, which means the communication between the tag and reader is successful. Thus we can affirm that the RF powering circuit can provide voltage source needed by the whole chip.

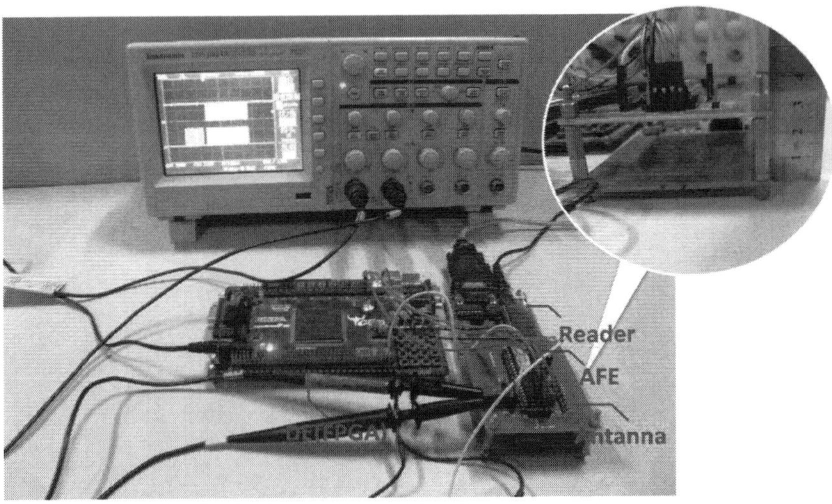

Figure 8. Functional measurement setup based on PCB and the demodulation signal from the tag.

The AFE was tested using a TDS 2024B oscilloscope. The test waveform of V_{cc} is shown in Figure 8. The strength of the electromagnetic field surrounding the RFID tag is adjusted by changing the distance between the chip and reader as shown in Figure 8. The V_{cc} test waveform when H = 0.3 A/m is shown in Figure

9a, and the one when H = 4 A/m is shown in Figure 9b. Other values of V_{CC} in different electromagnetic field are also tested, and they are shown in Figure 10. The test results of PCE of this RF powering circuit are shown in Figure 11.

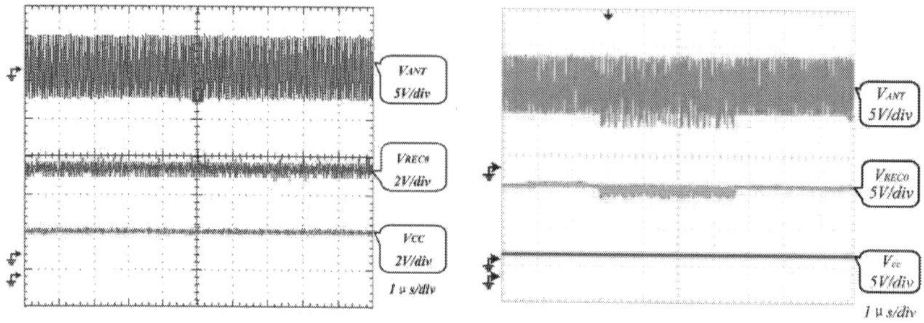

(a) (b)

Figure 9. (**a**) V_{CC} test waveform when H = 0.3 A/m under 0.25 μm process, V_{CC} = 2.40 V (rms); (**b**) V_{CC} test waveform when H = 4 A/m under 0.25 μm process, V_{CC} = 2.42 V (rms).

According to Figure 9, the Vcc voltage level is 2.4 V within the normal deviation ±10% of 2.5 V. According toFigure 10, considering the usual work environment of the tag, the regulator of this RF powering circuit has a quite small VRR with VRR = 2.5 mV/V while V_{RECO} is near 6 V. According to Figure 11, considering the usual work environment of the tag, in which the input power of the RF powering circuit is near 0.32 mW, the PCE of the RF powering circuit is 38.54%. Table 3 shows the performance comparison between this RF powering circuit and previous RF powering circuits.

Table 3. Comparison with previous RF power circuits.

	This Work	**[8]**	**[14]**	**[23]**	**[24]**
Process	0.25 μm	0.35 μm	0.18 μm	0.35 μm	0.35 μm
VRR (mV/V)	2.5	39	12	4	0.8
TRR (mV/°C)	0.04	Not Support	0.2	0.01	2

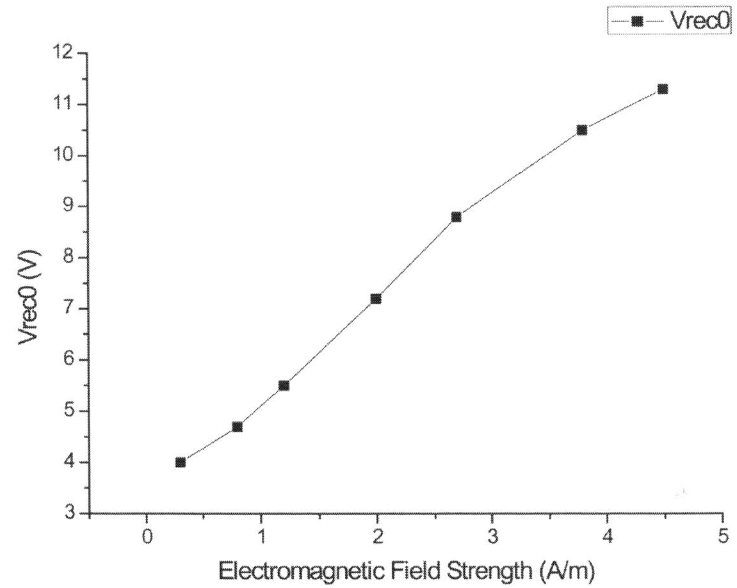

(a)

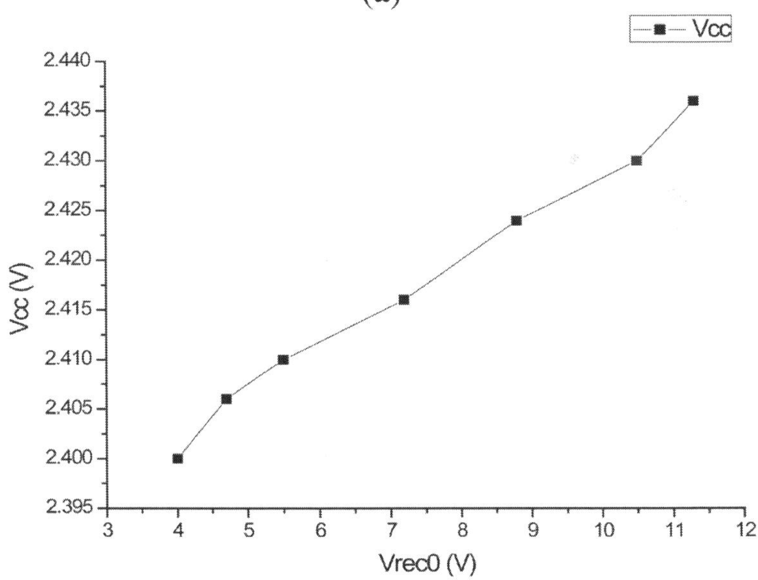

(b)

Figure 10. (**a**) V_{rec0} *vs*. Electromagnetic Field Strength; (**b**) V_{cc} *vs.* V_{rec0}.

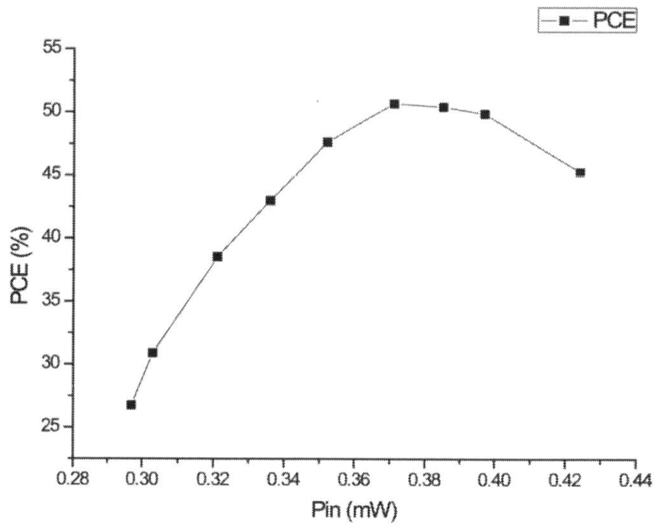

Figure 11. Test results of the PCE of the RF powering circuit in the HJTC 0.25 μm process.

Figure 12a shows that the demodulator of AFE has demodulated the downlink 10% ASK modulated signal sent from the reader successfully. The signal Demod10p is the demodulated signal achieved by the demodulator of AFE. Figure 12b shows an uplink signal of about 13.6% load modulated depth is achieved from the chip antenna to the reader which meets the communication criterion. The signal Datatran determined by the digital baseband controls the AFE modulator. Figure 13 shows that AFE has realized the information communication with reader successfully, which means this RF powering circuit work normally providing a stable voltage source for the whole chip.

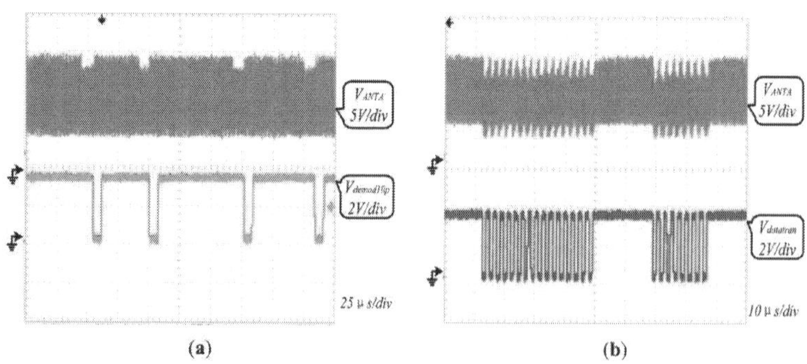

Figure 12. (**a**) Waveform of demodulation; (**b**) Waveform of modulation.

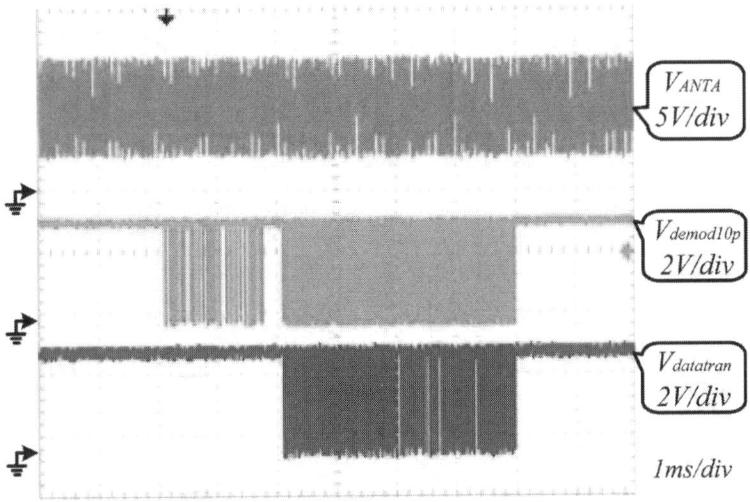

Figure 13. Measured waveform of the communication between reader and tag.

This RF powering circuit has been integrated in a RFID tag chip. The whole RFID tag packaged, chip die layout and chip photograph are shown in Figure 14. The size of the whole chip is 1.1 × 1.1 mm^2, whereas the RF powering circuit consumes 0.23 × 0.24 mm^2 of chip area. In Figure 14, there are also photographs of the packaged RFID tag and the tag used in a RFID smart card. The four pads for the package are two antenna pads and two voltage pads for testing.

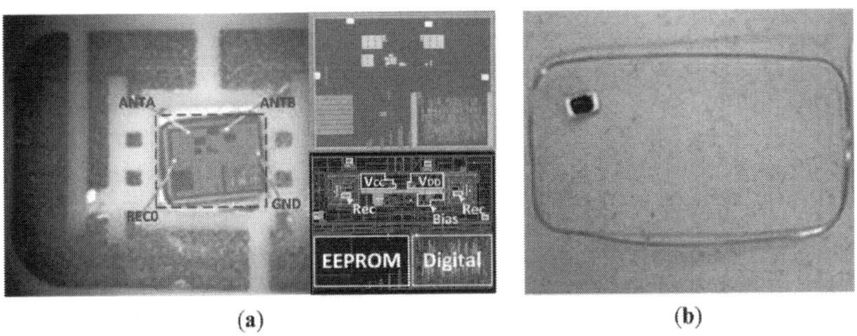

Figure 14. (**a**) Photograph of packaged RFID tag. Chip microphotograph and layout. (**b**) Chip tag used in a RFID smart card.

5. CONCLUSIONS

In this work, a RF Powering circuit used for passive RFID tags is designed and fabricated with the HJTC 0.25 μm CMOS process. The circuit consists of a rectifier, a bias circuit and a regulator. It provides two stable power sources for the analog and digital parts, respectively. The RF powering circuit achieves high PCE of 38.54% with the HJTC 0.25 μm process under normal working conditions and the power supply voltage is immune to temperature and voltage variations. Simulation results show that this circuit has a TRR = 0.04 mV/°C, and test results show that it has a VRR = 2.5 mV/V. This RF powering circuit is also fabricated with the HJTC 0.25 μm process in the ISO/IEC15693-Compatible and ISO/IEC 14443-Compatible RFID tag chip. Layout of this chip is optimized for latch-up prevention and noise isolation. Measurement results of the tag show that this circuit can provide a stable voltage source for a HF RFID tag chip for its proper operation.

ACKNOWLEDGMENTS

This work is supported by the National Natural Science Foundation of China (Grant No. 61306038 and No. 61376031), ChenGuang Project of science and technology for Wuhan youths (No. 2013072304010829) and the Fundamental Research Funds for the Central Universities (HUST: 2013TS050).

AUTHOR CONTRIBUTIONS

All authors have contributed to the presented work. Dongsheng Liu, Rencai Wang, Ke Yao, and Xuecheng Zou designed the whole circuit. Rencai Wang and Ke Yao conducted the implementation and wrote the manuscript. Liang Guo did the research on the regulator and the design of it. All authors have read and revised the manuscript.

REFERENCES

1. Chiao-Tzu, H.; Shu-Jen, W.; Wei-Ling, W.; Yung-Sheng, W. Construction of an Online RFID Enabled Supply Chain System Reliability Monitoring Model. Proceedings of the 2012 International Symposium on Computer, Consumer and Control (IS3C), Taichung, Taiwan, 4–6 June 2012; pp. 626–629.

2. Sangdo, P.; Hongchul, L. Self-Recognition of Vehicle Position Using UHF Passive RFID Tags. *IEEE Trans. Ind. Electron.* **2013**, *60*, 226–234.
3. Watfa, M.K.; Suleman, U.; Arafat, Z. RFID system implementation in Jebel Ali port. Proceedings of the 2013 IEEE Consumer Communications and Networking Conference (CCNC), Las Vegas, NV, USA, 11–14 January 2013; pp. 950–955.
4. Noorsal, E.; Sooksood, K.; Hongcheng, X.; Hornig, R.; Becker, J.; Ortmanns, M. A Neural Stimulator Frontend with High-Voltage Compliance and Programmable Pulse Shape for Epiretinal Implants. *IEEE J. Solid-State Circuits* **2012**, *47*, 244–256.
5. Cazeca, M.J.; Mead, J.; Chen, J.; Nagarajan, R. Passive wireless displacement sensor based on RFID technology. *Sens. Actuators A Phys.* **2013**, *190*, 197–202.
6. Rodriguez-Rodriguez, J.A.; Delgado-Restituto, M.; Masuch, J.; Rodriguez-Perez, A.; Alarcon, E.; Rodriguez-Vazquez, A. An Ultralow-Power Mixed-Signal Back End for Passive Sensor UHF RFID Transponders. *IEEE Trans. Ind. Electron.* **2012**, *59*, 1310–1322.
7. Roberts, N.E.; Wentzloff, D.D. A 98 nW wake-up radio for wireless body area networks. Proceedings of the 2012 IEEE Radio Frequency Integrated Circuits Symposium (RFIC), Montreal, QC, Canada, 17–19 June 2012; pp. 373–376.
8. Crepaldi, P.C.; Pimenta, T.C.; Moreno, R.L.; Rodriguez, E.C. A Low Power CMOS Voltage Regulator for a Wireless Blood Pressure Biosensor. *IEEE Trans. Instrum. Meas.* **2012**, *61*, 729–739.
9. Ferreira, P.M.; Bergeret, E.; Pannier, P. 1.4 V and 300 nA UHF passive RFID voltage regulator. Proceedings of the 2012 IEEE 10th International New Circuits and Systems Conference (NEWCAS), Montreal, QC, Canada, 17–20 June 2012; pp. 429–432.
10. DiGiampaolo, E.; Martinelli, F. A Passive UHF-RFID System for the Localization of an Indoor Autonomous Vehicle. *IEEE Trans. Ind. Electron.* **2012**, *59*, 3961–3970.
11. Chelho, C.; Young-Han, K.; Tae-Hun, K.; Kyusung, B.; Jongbae, K. Fully integrated ultra-low-power passive UHF RFID transponder IC. Proceedings of the 2011 IEEE International Symposium on Radio-Frequency Integration Technology (RFIT), Beijing, China, 30 November–2 December 2011; pp. 77–80.
12. Lijun, L.; Gannes, K.D.; Fricke, K.; Senjuti, S.; Sobot, R. Low-Power CMOS Voltage Regulator Architecture for Implantable RF Circuits. Proceedings of the 2012 Fourth International EURASIP Workshop on

RFID Technology (EURASIP RFID), Torino, Italy, 27–28 September 2012; pp. 99–106.

13. Jong-Wook, L.; Duong, H.T.V.; Quoc-Hung, H.; Sang-Hoon, H. A Fully Integrated HF-Band Passive RFID Tag IC Using 0.18-µm CMOS Technology for Low-Cost Security Applications. *IEEE Trans. Ind. Electron.* **2011**, *58*, 2531–2540.
14. Chi-En, L.; Yi-Jhan, H.; Jean-Fu, K. RFID Regulator Design Insensitive to Supply Voltage Ripple and Temperature Variation. *IEEE Trans. Circuits Syst. II Express Briefs* **2010**, *57*, 255–259.
15. Liu, D.; Li, F.; Zou, X.; Liu, Y.; Hui, X.; Tao, X. New Analysis and Design of a RF Rectifier for RFID and Implantable Devices. *Sensors* **2011**, *11*, 6494–6508.
16. Broutas, P.; Contopanagos, H.; Kyriakis-Bitzaros, E.D.; Tsoukalas, D.; Chatzandroulis, S. A low power RF harvester for a smart passive sensor tag with integrated antenna. *Sens. Actuators A Phys.* **2012**, *176*, 34–45.
17. Scorcioni, S.; Larcher, L.; Bertacchini, A. Optimized CMOS RF-DC converters for remote wireless powering of RFID applications. Proceedings of the 2012 IEEE International Conference on RFID (RFID), Orlando, FL, USA, 3–5 April 2012; pp. 47–53.
18. Qiuyun, F.; Gay, N.; Bansleben, C.; Hildebrand, R. Optimization of the NMOS and PMOS gate cross-connected rectifier for RF power extraction in RFID applications. Proceedings of the Asia-Pacific Microwave Conference (APMC 2008), Macau, China, 16–20 December 2008; pp. 1–4.
19. Gregoire, B.R.; Moon, U. Process-independent resistor temperature-coefficients using series/parallel and parallel/series composite resistors. Proceedings of the IEEE International Symposium on Circuits and Systems (ISCAS 2007), New Orleans, LA, USA, 27–30 May 2007; pp. 2826–2829.
20. Phillip, E.A.; Douglas, R.H. *CMOS Analog Circuit Design*, 2nd ed.; Publishing House of Electronics Industry: Beijing, China, 2011; pp. 109–124.
21. Momo, N.; Higashi, Y.; Oda, M.; Matsuzawa, K.; Kokubun, K.; Ohguro, T.; Momose, H.S.; Toyoshima, Y. 90 nm node RF CMOS technology with latch-up immunity on high-resistivity substrate. Proceedings of the European, Microwave Integrated Circuits Conference (EuMIC 2009), Rome, Italy, 28–29 September 2009; pp. 65–68.

22. Purwadi, A.; Shu-Ming, B.; Prabowo, B.A.; Jung-Ruey, T.; Gene, S. Shifting time waveform induced CMOS latch up in bootstrapping technique applications. Proceedings of the 2012 19th IEEE International Symposium on the Physical and Failure Analysis of Integrated Circuits (IPFA), Singapore, Singapore, 2–6 July 2012; pp. 1–4.
23. Minchin, L.; Chichun, H.; Zongwei, L. Implementation of Low Dropout Regulator with Low Bandgap Reference Voltage Circuit for RFID Tag Applications. Proceedings of the Cross Strait Quad-Regional Radio Science and Wireless Technology Conference (CSQRWC), New Taipei City, Taiwan, 23–27 July 2012; pp. 40–43.
24. De-Vita, G.; Iannaccone, G. Ultra-low-power series voltage regulator for passive RFID transponders with subthreshold logic. *Electron. Lett.* **2006**, *42*, 1350–1351.

Chapter 7

A Novel Integrated Circuit Driver for LED Lighting

Yanfeng Jiang, Dong Zhang, Ming Yu

Beijing Institute of Auto-Testing Technology, Beijing, China

ABSTRACT

A novel integrated circuit for driving LED lighting has been proposed, designed and fabricated. Besides the typical parts of LED driver, an integral part was added at the output terminal of error amplifier in the driver. In this way, a novel average current mode can be set up to take the place ordinary peak current control mode. In addition, a BUCK low-level topology was adopted, too. It can be used to drive up to eight 1 W HB LED lights with 350 mA constant current. In this way, the LED driver displays high performance, in which output current with less 1% error and total efficiency as high as 96%. The feasibility of the design has been verified by actual measurement on the fabricated chip.

Keywords: LED Lighting Driver, Integral Circuit, Low-Level Topology

1. INTRODUCTION

Contrasting to the traditional fluorescent lamp, high bright LED, being a potential lighting source, has the obvious advantage in energy saving, environmental protection, high efficiency, etc. [1] . So it has attracted much attention in academy and industry to push forward its utility. Based on the requirements of HB LED, some detailed limitations should be considered during the design of LED driver. For example, the life of LED will be dramatically decreased on the condition that there exists the driving current deviation [1] . So an ideal LED driver should have the constant output current with the high accuracy [1].

Currently, a peak current control mode has always been adopted in the existing LED driver to obtain constant current [2]. Based on statistical data [2], 30% error could be caused by this peak current control mode, which subsequently deteriorates LED life time dramatically, up to 20%.

The main reason for influencing the accuracy of output current at peak current control mode lies in the fact that the peak current is varied with the inductor in the driver circuit [3]. Here, an average current control mode was adopted, in this case it isn't influenced by the inductor value anymore. In this way, a higher accurate output will be obtained in LED driver to guarantee its lifetime.

Besides the above mentioned accuracy problem, power efficiency is also a key parameter to evaluate the driver's quality. The topology of driver circuit will determine the efficiency [3]. In this paper, an optimized topology will be used to reduce power consumption.

To solve the above issues, an integrated circuit driver with average current control mode was proposed and designed in this paper, to get accurate and stable output current, which will be introduced in the Section 2. A BUCK low-level topology will be adopted in the driver to get high efficiency. This will be introduced in Section 3. Section 4 shows the simulation results using CSMC 0.5 µm 40 V BCD model. Finally, the experimental results and discussion will be made to show design's advantage.

2. CHIP DESIGN WITH AVERAGE CURRENT CONTROL MODE

Traditionally, in order to reduce the cost and complexity of the chip, a peak current mode control mode is always used in LED driver chip [4] [5]. However, the limitation of this method is the output current is easily influenced by inductor L and input voltage. This current variation will lead to degeneration of accuracy. At the same time, the actual current provided for the HB LED is less than the peak current. For this mode, theoretically speaking, the value of inductor should be infinity to obtain an output current which is equal to the design value.

Figure 1 schematically shows the current in the peak current control mode in the common BUCK topology. It can be seen from this figure that different inductor corresponding to same peak current, but different average currents. The greater the inductance L is, the closer the average approaches to peak current. Because of practical limitations, the inductance value L can not large enough. So the actual current value flowing through the LED is always less than the peak current. Based on the above analysis, there are two shortcomings for the peak current control mode. The one is that its average current is always varied by the inductor and input voltage. It can't maintain stable all the time. The other shortcoming lies in the fact that the average current is different from the designed peak current. These two facts have an obvious influence on the output's accuracy.

Figure 2 shows the schematic of the average current control mode. It can be seen that the average current always maintain the same regardless of inductor L values. Comparing the two control modes shown in Figure 1 and Figure 2, one conclusion can be made that the average current control mode can be an effective method to guarantee the accuracy of output current. So, in this work, the average current control mode will be fulfilled in the circuit design. An integrated circuit with this mode has been designed, which schematic is shown in Figure 3. The part within dashed line corresponds to the designed integrated circuit, which includes error amplifier (EA), pulse-width modulator (PWM), slope generator, RS flip-flop, output stage with power transistor, etc. Its structure is a typical PWM controller except there is an integral part at the output terminal of EA.

The integral part is composed by three components, c_1, c_2 and R_2. For the two input terminal of EA, positive one is connected to a voltage reference while negative one $V_{sense}(t)$, termed as the detection voltage, connected to outside sense resistor, as shown in Figure 3. When the power transistor is on-state, there will be a current flowing through the sense transistor, so the detection voltage will appear and moreover, it will be increased for the existence of an external inductor L, which is not displayed in this figure.

At beginning, detection voltage is smaller than reference, EA will have a positive output. Because the integral parts exist, capacitors c_1, c_2 will be charged so that the voltage at output terminal increases.

When the value of detection voltage is larger than the reference, output of EA should be negative. However, the integral part will try to maintain the original EA output voltage. So, there is a discharge phase during this period. There are two main work states: In the first state, when power transistor is on-state, the detection voltage will appear, increasing during the on-state, as shown inFigure 4(a). The output of EA, $V_{EA}(t)$, as shown in Figure 4(b), rises at the beginning. When the detection voltage equals to the reference, $V_{EA}(t)$ reaches to the maximum value. Afterwards, with detection voltage increasing, output of EA will be decreased.

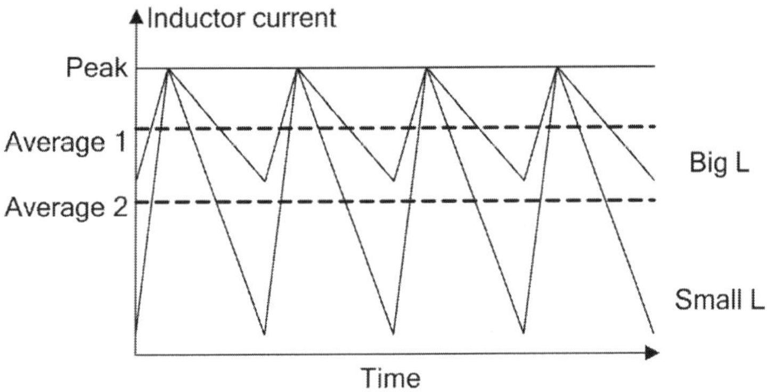

Figure 1. Schematic of peak current control mode.

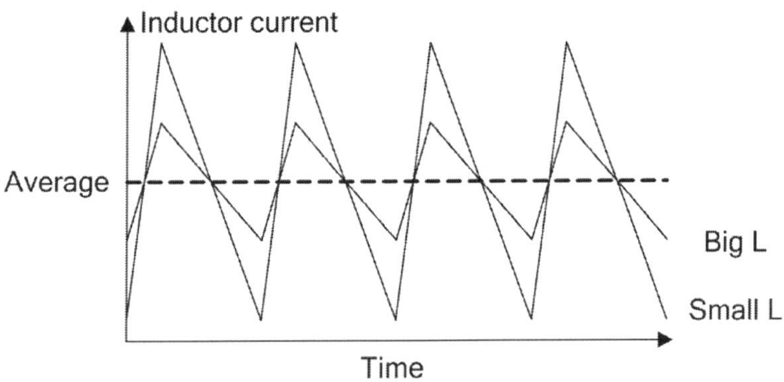

Figure 2. Schematic of average current control mode.

In the second state, when N-MOSFET power transistor turns off, $V_{EA}(t)$ remains unchanged for the existence of integral circuit, as shown in Figure 4(b). Figure 4(c) shows the waveform of slope, which is connected to PWM's negative input terminal, denoted as $V_{slope}(t)$. $V_{EA}(t)$ is connected to positive ones. The average value of detection voltage during on-state has been set to be equal to the reference voltage. Figure 4(d) shows the output of PWM modulator. So the system is operated in a fixed period of conduction control mode. As is shown in the dotted line area in Figure 3, it is the core circuit for the average current regulation mode.

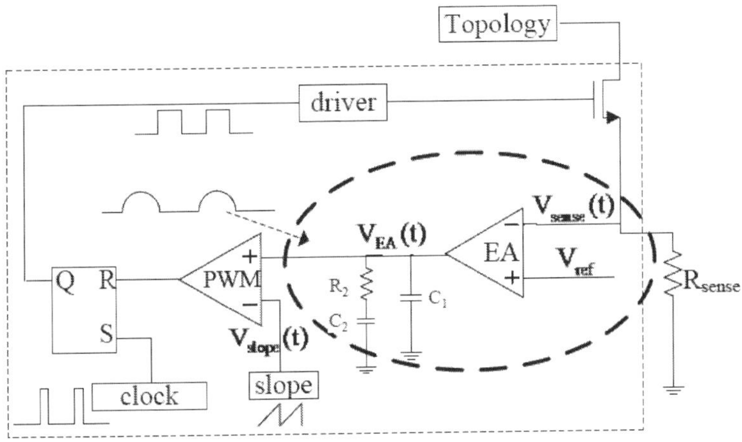

Figure 3. Schematic diagram of driving circuit.

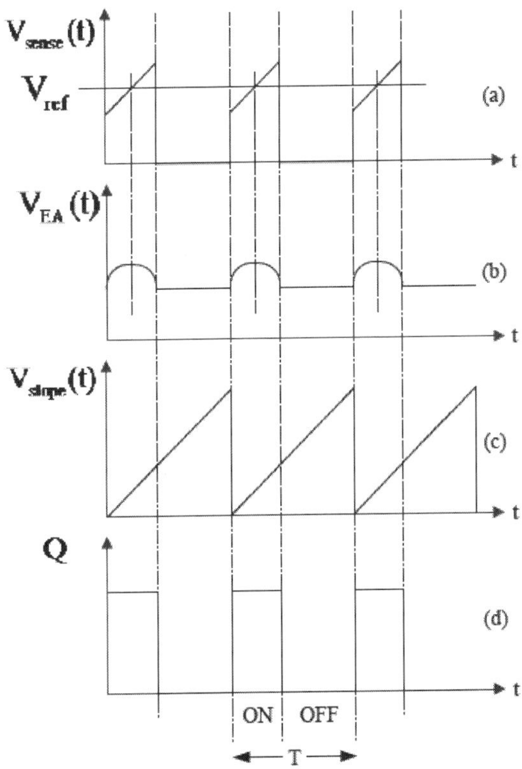

Figure 4. Timing diagram of driving circuit.

The traditional current regulator was designed as peak current control mode. It can't provide the actual average current regulation for the LED lamps. The traditional circuit uses an error amplifier to compare the difference between the transient value of detection voltage $V_{sense}(t)$ and the reference voltage. For the circuit shown in Figure 3, which is the average current mode and carries out the time integration about the error, the capacitors take the integration of the difference between the detection voltage $V_{sense}(t)$ and the reference voltage V_{ref}. Because the reference voltage V_{ref} is a constant value, that is to say, during the integration time $T * D$ (D is the switching duty cycle, T is the period time), it enables the difference between the average of $V_{sense}(t)$ and the reference voltage drop to a minimum value, so:

$$\int_0^{TD}\left[V_{ref}-V_{sense}(t)\right]dt = V_{ref} * TD - \int_0^{TD} V_{sense}(t)\,dt \qquad (1)$$

Supposing Δv as the capacitance variation during Δt period, that is the increment of $V_{EA}(t)$. The equivalent output capacitance of EA is c, and the transconductance of EA is g_m. Following equation can be obtained:

$$\begin{aligned}\Delta V_{EA}(t) = \Delta v &= \frac{i\Delta t}{c}\\ &= \frac{\Delta t}{c} g_m\left[V_{ref}-V_{sense}(t)\right]\\ &= \frac{g_m}{c}\left[V_{ref}-V_{sense}(t)\right]\Delta t\end{aligned} \qquad (2)$$

Based on Equation (2), when V_{ref} equals to $V_{sense}(t)$, the voltage on the capacitor reaches maximum value. When V_{ref} is less than $V_{sense}(t)$, the capacitor discharges, and then when V_{ref} is larger than $V_{sense}(t)$, the capacitor charges. So, during the TD period, $V_{EA}(t)$ increases at first. Then, it will decrease until the capacitor voltage equals to its initial value. When the state changes, the value of $V_{EA}(t)$ will remain the same during $T * (1-D)$ period. In this way, it repeats every cycle to form a stable average current.

It can also be noted that the output of EA is composed of a resistor R_2 and two capacitors c_1 and c_2, which forms two poles and one zero, and its frequency domain characteristics can be expressed as:

$$H(jw)=\frac{1+j\omega c_2\times R_2}{j\omega(c_1+c_2)+(j\omega)^2 c_1\times c_2\times R_2} \tag{3}$$

It can be seen that the corresponding zero point is $\frac{1}{c_2\times R_2}$ and the poles are 0 and $-\frac{c_1+c_2}{c_1\times c_2\times R_2}$. The zero can be inserted to eliminate the poor phase margin, which caused by the second pole, to increase the stability of the circuit.

3. IMPROVED TOPOLOGY

The common non-isolated BUCK topology is shown in Figure 5. V_{IN}, as the input voltage, is applied on drain terminal of power transistor. Based on the detection voltage of feedback resistor R, the power transistor is used to adjust conduction time.

In addition, the power transistor in Figure 5 is located at high-level, and it adopts gate drive technology of the bootstrap circuit, ignoring the loss of power transistor. When V_{IN} is relatively high, gate voltage should be at least V_{IN} + V_{TH}. The transistor has high risk of breakdown. Compared with the topology shown in Figure 5, an improvement has been made in Figure 6, in which the power transistor is placed at low level. In this way, the low level N-MOSFET is much safer than high level ones.

As shown in Figure 5, in the traditional Buck topology, considering some key problem as the heat dissipation, maintenance or the module replacement, most components in controller are separated from HB LED. The current sense resistor R will be placed near the ground, and it only needs single-end detection circuit instead of complex differential detection circuit. So it can prevent electromagnetic interference.

In the system shown in Figure 6, the sensor resistor R is placed between source terminal and ground. Its power efficiency is increased compared to the circuit shown in Figure 5 [6]. In addition, in Figure 6, the lowlevel N-MOSFET and sense resistor R can transmit inductive current only during part of the cycle, but the power loss of R in Figure 5 occurs during the entire cycle, so the power loss of R in Figure 6 equals to that of R in

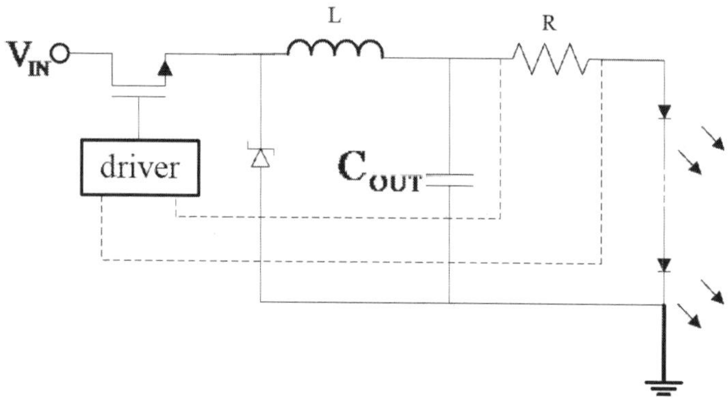

Figure 5. Traditional BUCK topology.

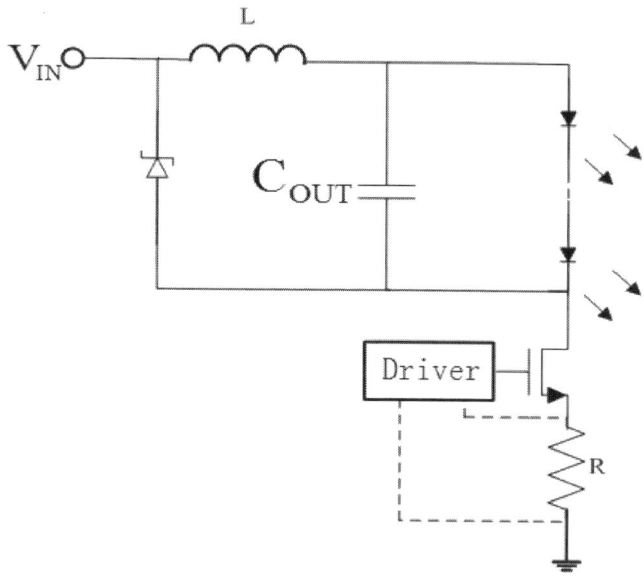

Figure 6. Optimized structure of LED driver.

Figure 5 multiplied by the switch duty cycle D, and the value of D is usually lower than one. Therefore, in Figure 6, the power loss of R will be reduced during a switching period $(1-D)T$, which can be shown as:

$$P=\frac{1}{T}\int_0^{DT} i_{\mathrm{LED}}^2(t)R\,dt \tag{4}$$

In Figure 5, the power loss of R can be expressed as:

$$P' = \frac{1}{T}\int_0^T i_{\mathrm{LED}}^2(t)R\mathrm{d}t \tag{5}$$

The saved power can be expressed as:

$$\Delta P = \frac{1}{T}\int_0^{(1-D)T} i_{\mathrm{LED}}^2(t)R\mathrm{d}t \tag{6}$$

According to the characteristic of the inductor, the above formula can be further shown as:

$$\begin{aligned}\Delta P &= \frac{1}{T}\int_0^{(1-D)T}\left(I_{\mathrm{peak}} - \frac{V_L}{L}t\right)^2 R\mathrm{d}t \\ &= R(1-D)\left[I_{\mathrm{peak}}^2 - \frac{V_L I_{\mathrm{peak}}}{L}T(1-D) + \frac{1}{3}\frac{V_L^2}{L}T^2(1-D)^2\right]\end{aligned} \tag{7}$$

where T is the switching cycle, R is the sense resistor, and I_{peak} indicates the peak current of the inductor L, and V_L is the voltage across the inductor L. When the transistor is off-state, during $(1-D)T$, V_L equals the voltage drop across the LED lights. It can be seen from the above equation that placing the sense resistor on the low side will reduce the power loss.

The schematic shown in Figure 7 is a complete LED driver circuit structure. It uses the average current control mode as shown in Figure 3, which increases the accuracy of the output current. At the same time, the optimized topology as shown in Figure 6 will reduce the overall power loss.

4. SIMULATION RESULT AND LAYOUT DESIGN

This design is implemented in CSMC 0.5 μm 40 V BCD technology. Figure 8 is the simulation result of gain and phase margin of error amplifier. By the reasonable settings of c_1, c_2, R_2 in Figure 3 to adjust the location of zero, the stability of the circuit is enhanced. From the diagram it can be seen that the gain of the error amplifier is 60 dB, and the phase margin, which ranges from 180 to 47.75 degrees, is about 132 degree.

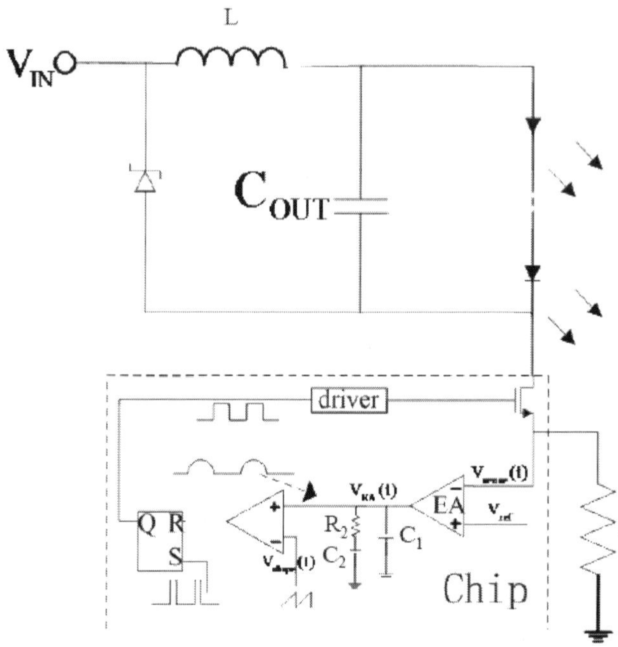

Figure 7. Complete LED driver circuit structure, the part shown within dashed line denotes the designed integrated circuit.

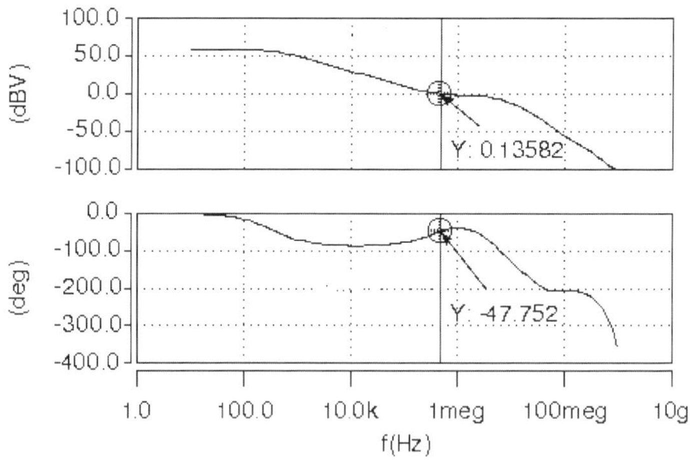

Figure 8. Bode plot of the core circuit.

Figure 9 shows the driving current curve through the LED lamp. Based on the BUCK topology inFigure 6, a suitable inductor L should make the ripple current amplitude as small as possible. The greater inductor will produce more accurate output current. However, in order to minimize the physical size of the circuit, the value of inductor should be selected to enable the circuit working in continuous conduction mode. At the same time, it should be guaranteed that the peak current of the inductor does not exceed its saturation current. Calculation result shows that the value of L is over 22 μH. As shown in Figure 9, in which L equals to be 22 μH, C_{out} equals 10 μF, V_{in} equals 32 V, and eight 1 W of HB LED lights acting as load. It can be seen from Figure 9 that the ripple of transient output current is 4.8 mA and the error is about 0.68%, indicating that the output current is stable and has high accuracy.

Figure 10 corresponds to the situation as same as that in Figure 9 except that L equals 44 μH. The ripple of the transient output current is 1.6 mA, the error is about 0.23% with unchanged average current.

Based on the values shown in Figure 9 and Figure 10, although the difference of the inductance value is twice, its average current remains constant, and the ripples are small. For the traditional peak current mode, its average current and ripple depend heavily on the value of inductor L. So, the average current mode has good stability to the output current, not changing with the value of the inductor L.

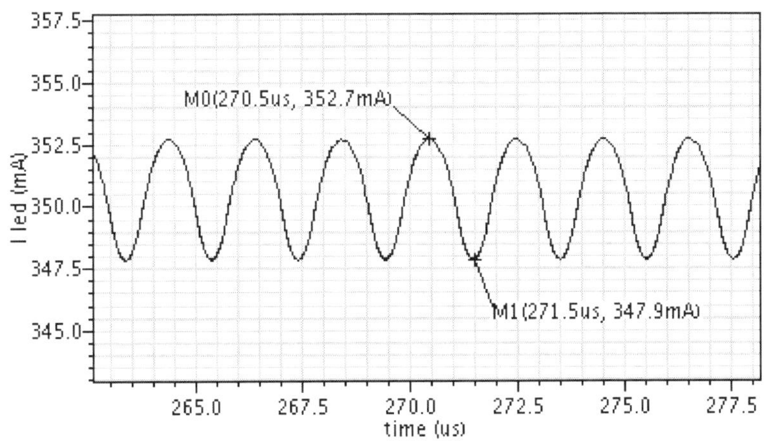

Figure 9. Output drive current ripple when L = 22 μH.

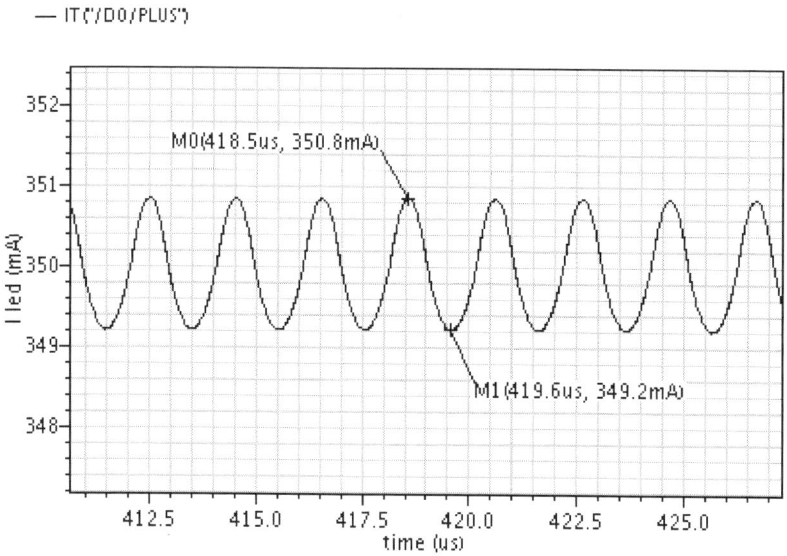

Figure 10. Output drive current ripple when L = 44 μH.

Figure 11 shows the efficiency of HB LED driver circuit. Taking into account the heat dissipation problem of HB LED lights, the simulation curve displays the efficiency when the ambient temperature is 90 degrees celsius. The abscissa denotes the input voltage V_{in}, and the vertical axis is the efficiency. Three curves corresponds different loads from top to bottom, which are eight LED lights, six LED lights and two LED lights. It can be seen that all efficiency are above 91% and the highest is above 96%.

Figure 12 shows the layout of this design. Because the power transistor is integrated into the chip, its power consumption is converted into heat during normal working. During the chip layout, the power MOS transistor should be placed at the edge of chip and an isolation belt has been placed to decrease interference to other circuits. The power supplies of digital and analog circuits should be separated. Considering its larger transient voltage and current in digital circuit, in order to reduce its interference in analog circuits, it is necessary to separate the power supply. It is also paid attention to the total match of difference pair, and the match between resistors and capacitors and so on. The final layout size is 1.4 mm × 1.4 mm.

Using the LED driver chips used in the external topology, the system is verified, mainly focusing on the accuracy of output current and the efficiency at different L values, varied from 22 μH to 55 μH. The number of LED lights remains eight all the time while V_{in} remains 32 V and C_{out} is fixed to be 10 μF. The measured result is shown in Figure 13. The output currents remain almost same under different inductors, totally within 0.3% error.

Table 1 shows the test results of the average current and efficiency, in which L = 22 μH, C_{out} = 10 μF, V_{in} = 32 V, and the numbers of 1 W LED lights are 2, 6, 8, respectively.

The average currents, corresponding to different number of LED, are larger than expecation value. The error is less than 1%, which may be caused by the process variation and thermal factors, etc.

The efficiency is 96% with 8 LED lights, 95% with 6 lights and 92% with 2 lights, respectively. The data shows the high efficiency of the designed integrated circuit and the application topology.

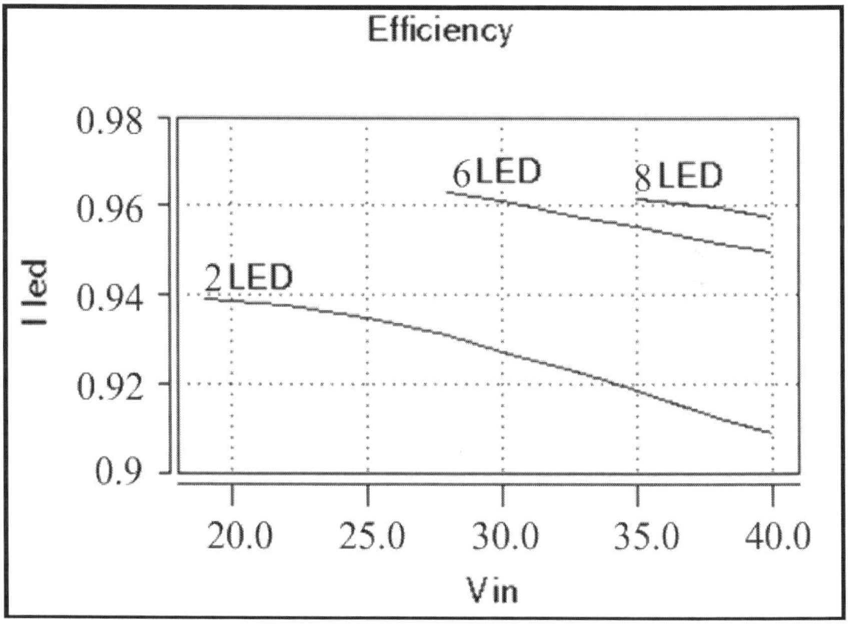

Figure 11. The efficiency of HB LED driver circuit.

Figure 12. Layout design of the chip.

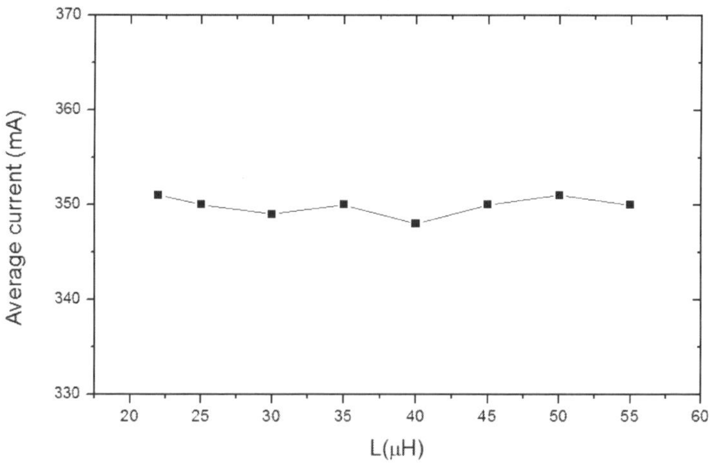

Figure 13. Relationship between average current and the inductance L.

Table 1. Main parameters of the test results.

Test parameters	Num. of LED lights	Size	Unit
Average current	2	352	mA
	6	351.5	
	8	350.6	
Efficiency	2	92	%
	6	95	
	8	96	

5. CONCLUSION

A LED lighting driver has been designed, including an integrated circuit with integral part and the buck converter topology with low-level switch and low-side sense resistor. The integrated circuit has been fabricated based on the CSMC 0.5 μm 40 V BCD process. In this way, the driver with high accurate output current and high efficiency can be obtained. The final result shows that the output current error is less than 1% and the efficiency is above 96%. Since a significant improvement has been obtained, the circuit is potentially applicable for future LED lighting.

ACKNOWLEDGEMENTS

This project is supported by Beijing Natural Science Foundation (4122031) and Natural Science Foundation of China (60876078).

REFERENCES

1. Leung, W.Y., Man, T.Y. and Chan, M. (2008) A High-Power-LED Driver with Power-Efficient LED Current Sensing Circuit. Proceedingsof Solid-State Circuits Conference, Edinburgh, 15-19 September 2008, 354-357.
2. Ma, F.F., Chen, W.Z. and Wu, J.C. (2007) A Monolithic Current-Mode Buck Converter with Advanced Control and Protection Circuits. IEEE Transactions on Power Electronics, 22, 1836-1846.
3. van Heinz, B., Georg, S. and Matthias, W. (2007) Power Driver Topologies and Control Schemes for LEDs. APEC 2007—Twenty Second Annual IEEE, Anaheim, 25 February-1 March 2007, 1319-1325.

4. Deng, H., Duan, X., Sun, N., Ma, Y., Huang, A.Q. and Chen, D. (2005) Monolithically Integrated Boost Converter Based on 0.5-μm CMOS Process. IEEE Transactions on Power Electronics, 20, 628-638.
5. Lee, C.F. and Mok, P.K.T. (2004) A Monolithic Current-Mode CMOS DC-DC Converter with On-Chip Current-Sensing Technique. IEEE Journal of Solid-State Circuits, 39, 3-14.
6. Benini, L., Bogliolo, A. and De Micheli, G. (2000) A Survey of Design Techniques for System-Level Dynamic Power Management. IEEE Transactions on Very Large Scale Integration Systems, 8, 299-316.

Chapter 8

A Multi-Channel Low-Power System-on-Chip for in Vivo Recording and Wireless Transmission of Neural Spikes

Andrea Bonfanti [1,*], Maria Ceravolo [1,†], Guido Zambra [1,†], Riccardo Gusmeroli [1], Gytis Baranauskas [2], Gian Nicola Angotzi [2], Alessandro Vato [2], Emma Maggiolini [2], Marianna Semprini [2], Alessandro Sottocornola Spinelli [1] and Andrea Leonardo Lacaita [1]

[1]Dipartimento di Elettronica e Informazione, Politecnico di Milano, Via Golgi 40, Milano 20133, Italy

[2]Italian Institute of Technology (IIT), Department of Robotic, Brain and Cognitive Science, Genova 16163, Italy

[†]Present address: STMicroelectronics, Via Tolomeo, Cornaredo 20010, Italy.

ABSTRACT

This paper reports a multi-channel neural spike recording system-on-chip with digital data compression and wireless telemetry. The circuit consists of 16 amplifiers, an analog time-division multiplexer, a single 8 bit analog-to-digital converter, a digital signal compression unit and a wireless transmitter. Although only 16 amplifiers are integrated in our current die version, the whole system is designed to work with 64, demonstrating the feasibility of a digital processing and narrowband wireless transmission of 64 neural recording channels. Compression of the raw data is achieved by detecting the action potentials (APs) and storing 20 samples for each spike waveform. This compression method retains sufficiently high data quality to allow for single neuron identification (spike sorting). The 400 MHz transmitter employs a Manchester-Coded Frequency Shift Keying (MC-FSK) modulator with low modulation index. In this way, a 1.25 Mbit/s data rate is delivered within a limited band of about 3 MHz. The chip is realized in a 0.35 μm AMS CMOS process featuring a 3 V power supply with an area of 3.1 × 2.7 mm^2. The achieved transmission range is over 10 m with an overall power consumption for 64 channels of 17.2 mW. This figure translates into a power budget of 269 μW per channel, in line with published results but allowing a larger transmission distance and more efficient bandwidth occupation of the wireless link. The integrated circuit was mounted on a small and light board to be used during neuroscience experiments with freely-behaving rats. Powered by 2 AAA batteries, the system can continuously work for more than 100 hours allowing for long-lasting neural spike recordings.

Keywords: neural recording; action potential; low-noise amplifier; wireless transmitter; bio-telemetry; FSK; Manchester-encoding

1. INTRODUCTION

There is a growing need for wireless implantable neural recording systems that can simultaneously acquire neural signals from a large number of channels. Recording from hundreds of neurons may allow neurophysiologists to study brain function in freely-behaving animals, and in future will provide the signals necessary to control neural prosthetic devices [1]. Compared to wired systems, wireless neural recording devices provide a number of benefits: reduction of motion artifacts and freedom of animal movements in neuroscience experiments, as well as reduced infection risks and higher patient mobility in neuroprosthetic applications. However, these systems entail severe technical challenges, such as the robustness of the wireless link and the need of a large data transmission band, coupled to low-power dissipation and small size in case of brain-implantable devices. The main issue is related to the huge amount of data associated with a large number of recording channels: in a 64 channel system featuring a 20 kHz sampling frequency (the neural signal band extends to about 7–10 kHz) with 8 bit resolution, the required data rate of the wireless link is 10.24 Mbit/s. Achieving such a high bit rate is not straightforward either in fully implantable devices where power consumption is limited to about 10 mW to avoid tissue necrosis [2] and in battery-powered systems for maintaining a reasonable battery life. Hence, most integrated circuits (ICs) reported in the literature that simultaneously support neural recording and wireless telemetry must trade-off power consumption against the amount of data to be wirelessly sent.

Data compression was adopted to solve this problem in earlier systems. For example, since in a typical extracellular electrode trace most of the useful information is contained in spikes (or action potentials, APs), Harrison *et al.* [3] enable only the transmission of the spike occurrence information, achieving a low bit rate of 330 kbit/s for 100 channels. However, since the examination of the raw traces may improve the selection of the spike detection threshold, the system reported in [4] allows to collect raw data from 2 out of 64 channels present in the system, while only spike occurrence times are sent for the remaining 62 channels via a 2 Mbit/s wireless link. On the other hand, spike waveforms are used in neuroprosthetic applications to assign spikes to specific neurons improving the prediction of an intended movement [5]. Hence, during the last four years, wireless systems capable of sending all raw data have been reported. In [6], Chae *et al.* describe a 128 channel system able to transmit all raw data using an impulse-radio ultra-wide band (IR-UWB) transmitter with up to 90 Mbit/s rate. In such system, problems may arise from skull and skin absorption at high frequencies (signal spectrum is in the 3.5–4.5 GHz range) and from pulse distortion and the need of an external broadband antenna. As a matter of fact, no *in vivo* tests are reported for this kind of system. In [7], Ghovanloo *et al.* report a system that transmits raw data from 32 channels, encoded in a pulse-width modulated (PWM) signal via a 640 kS/s frequency-shift keying (FSK) wireless link. In this system, the transmitted information relies on the time width of each symbol, making this transmission prone to the noise of the wireless link and of the receiver. Moreover, the system in [7] requires a relatively large bandwidth (38 MHz), which means higher probability of RF interferences, and does not meet the band requirements of any medical communication standard.

We propose here a solution that is intermediate to the two extremes described above. The idea was to employ a data compression algorithm in order to:

- preserve the information needed for single neuron identification;
- reduce the throughput and thus the power consumption, since the power needed to the antenna is directly proportional to the bit rate [8];
- keep the bandwidth limited to few MHz, thus reducing the probability of RF interference and verifying the possibility to make in the future

the system compliant to Medical Implanted and Communication Service (MICS) or Industrial Scientific and Medical (ISM) bands, in the 402–405 MHz and 902–928 MHz frequency range, respectively.

In particular, having reduced the data rate, it was possible to extend the transmission distance well beyond the 1 m value reported in the above-mentioned works to make possible realistic *in vivo* experiments with a small penalty in terms of power consumption. The proposed integrated circuit, already presented in [9], has been included in a complete system, made of a transmission and a receiving unit, that is in use in a neuroscience laboratory on freely-behaving animals in place of commercial wired systems to avoid tethering effects and to provide electrical isolation. In this paper we describes in detail the architecture and circuit design choices, the electrical performance of the IC and the *in vivo* tests on small-laboratory animals.

The paper is organized as follows. Section 2 describes the overall system made of a signal conditioning and transmitting unit and a receiver built with off-the-shelf components. Section 3 discusses in detail the implementation of the system-on-chip (SoC) building blocks and validates the data compression algorithm by means of ad-hoc simulations, while Section 4 shows the electrical characterization of the integrated circuit and reports the results of *in vivo* tests on freely-behaving laboratory animals. Finally, a comparison of the implemented SoC with the state-of-the-art is presented in Section 5.

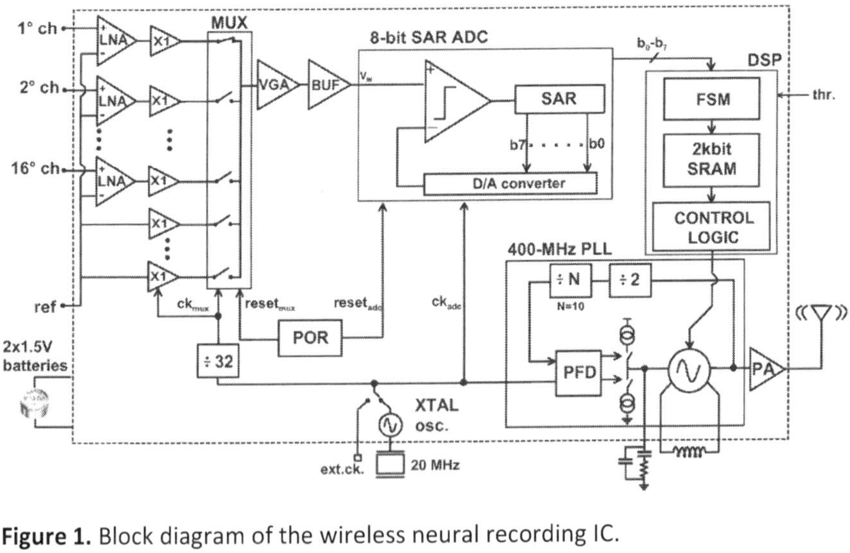

Figure 1. Block diagram of the wireless neural recording IC.

2. SYSTEM ARCHITECTURE

The system consists of a wireless recording unit, a receiver built with off-the-shelf modules to maximize its sensitivity plus a remote host complete of a graphical user interface (GUI) to allow neural signal visualization during *in vivo* experiments.

2.1. Wireless Recording Unit

Figure 1 shows the block diagram of the IC which acts as recording and transmitting module. The circuit consists of 16 low-noise amplifiers (LNAs), which perform low-power and low-noise amplification of neural signals. The amplifier outputs are sampled by a time-division multiplexer (TDM) at 20 kS/s onto one data lead before being further amplified by a variable-gain amplifier (VGA) and converted by an 8 bit successive approximation register (SAR) analog to digital converter (ADC). The digital data are then processed by a logic block, performing data compression, assembly and Manchester encoding of the bit stream. Finally, the digital signal is sent to a 400 MHz binary FSK wireless transmitter based on direct modulation of a Voltage Controlled Oscillator (VCO), enabling a 1.25 Mbit/s data rate. A crystal Clapp oscillator employing an external 20 MHz quartz and a power-on/reset (POR) circuit complete the system, providing a clean 20 MHz frequency clock and a synchronization signal at the start-up. It is worth pointing out that, to reduce silicon area and costs, only 16 amplifiers have been integrated. The remaining 48 channels have been added by connecting a voltage reference to the multiplexer input. In this way all the stages following the amplifier array are designed to process signals from a 64 channel front-end, thus making it possible to reliably assess the actual performance of a complete 64 channel system.

2.2. Receiver and Graphical User Interface

The receiver (see Figure 2) consists of a quadrature zero-IF down-converter (MAX3580) using an LNA with a noise figure of 4.7 dB at 400 MHz, RF and baseband tunable filters and automatic gain control circuits. The quadrature signals are converted by a dual channel, 20 MSps, 10 bit ADC (AD9201) and fed to a Xilinx FPGA module (OpalKelly, XEM3005), which performs frequency demodulation and Manchester decoding and sends the received data to a PC via a USB link. In addition, a Graphical User Interface (GUI) based on Labview/C software was implemented to allow data saving and on-line elaboration.

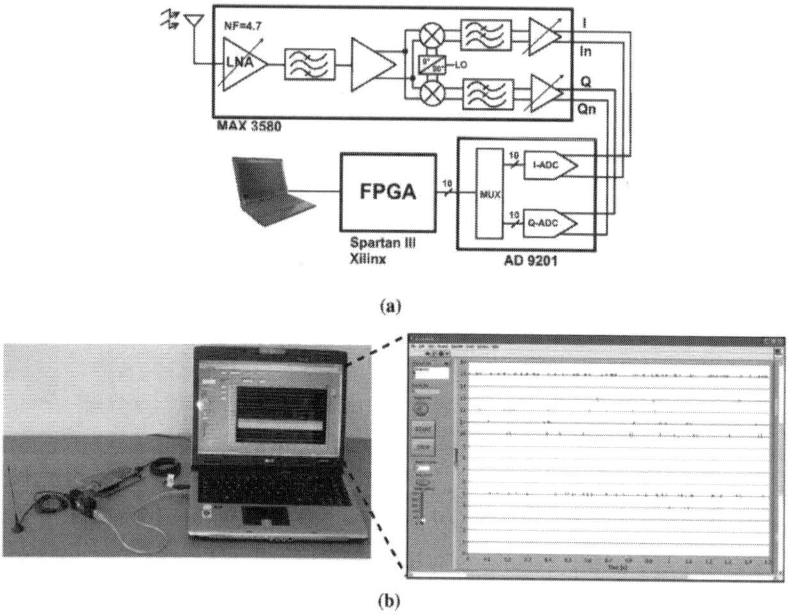

Figure 2. Block diagram of the receiver (**a**) and photo of the remote receiver with the detail of the graphical user interface (**b**).

3. CIRCUIT DESIGN

3.1. Analog Front-End

Amplification and filtering of the input signal are performed by a three-stage circuit shown in Figure 3. The first stage is an AC-coupled high-pass filter, using two MOS-bipolar pseudo-resistors as feedback elements [10], which enables the synthesis of high-value resistances without using large-area components. The mid-band amplifier gain is given by $G_1 \cong -C_1/C_2$, set to –67 by taking $C_1 = 10$ pF and C_2 = 150 fF. The high-pass pole frequency is placed below 10 Hz to reject the offset and the slow voltage drift of the electrode. A tunable G_M-Chigh-pass filter with a cut-off frequency f_{HP} of about 300 Hz is introduced after the first stage. It is designed to reject the low-frequency signals, such as Local Field Potentials (LFPs) in the $1-100$ Hz frequency range that can prevent a correct detection of the neural spikes or even saturate the amplifier, and cuts off the input-referred $1/f^2$ noise due to the pseudo-resistors of the first amplification stage. After the selective high-pass filter, a second non-inverting gain stage is added to provide further signal amplification and to define the high frequency cut-off, f_{LP}. The stage is a single-ended capacitive-coupled voltage amplifier with a gain $G_2 \approx 33$, achieved by using $C_3 = 5$ pF and $C_4 = 150$ fF. A capacitive-coupled structure was preferred to a purely resistive feedback amplifier to minimize the current drawn by the OTA output stage. The DC voltage at the amplifier input is determined by the pseudo-resistor elements in the feedback path, while the low-pass cut-off frequency is set by the gain-bandwidth product of the operational amplifier ($GBWP_2$) to about $GBWP_2/G_2 \cong 10$ kHz. Note that at very low frequencies this non-inverting stage has a unity gain in order not to amplify the offset of the operational amplifier.

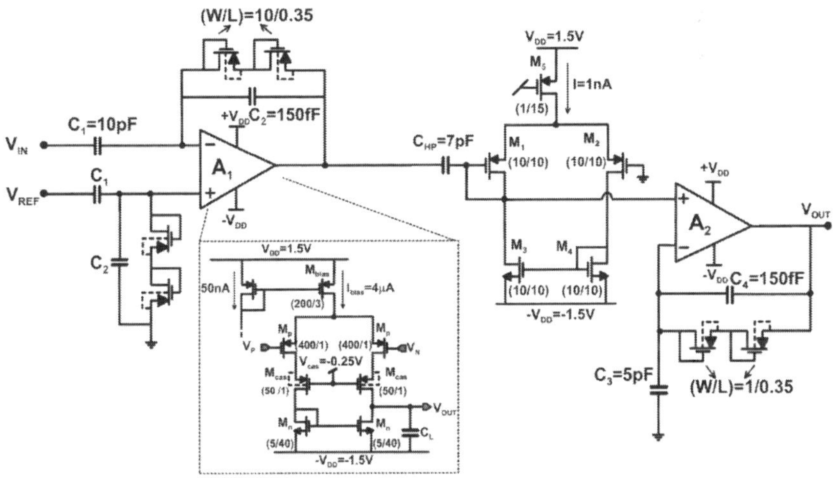

Figure 3. Schematic of the low-noise band-pass neural pre-amplifier.

3.1.1. Noise Analysis and First Stage Sizing

In order to minimize the amplifier array power consumption while maintaining the input-referred noise specifications, a rigorous noise analysis of the preamplifier and filter has to be carried out following the approach adopted in [10]. The noise performance of the whole LNA depends on the design of the first-stage OTA, whose input noise power density is mainly determined by a compromise between its thermal and flicker noise. To explain this point, let us denote as C_p the input parasitic capacitance to ground of each input terminals of the first-stage OTA [11]. If we neglect the noise contribution due to the two pseudo-resistors (this term will be addressed in a next step), we can write the input-referred noise of the overall amplifier as:

$$\overline{E^2}_{n_{eq}} = \left(\frac{C_p + C_1 + C_2}{C_1}\right)^2 \overline{E^2}_{n_{eq,OTA}} = \left(1 + \frac{1}{|G_1|} + \frac{C_p}{C_1}\right)^2 \overline{E^2}_{n_{eq,OTA}} \tag{1}$$

where $\overline{E^2}_{n_{eq,OTA}}$ is the input-referred voltage noise of the first-stage OTA. Equation (1) highlights that C_p, which is mainly determined by the input transistor dimensions, cannot be increased too much without compromising the overall input-referred noise. On the other hand, it is well known that large-area input transistors are required to minimize the OTA flicker noise. Therefore, to achieve the best overall performance these two contradictory requirements have to be balanced. Furthermore, according to Equation (1), the first stage gain, G_1, can be increased maximizing C_1. Its value is limited by the electrode impedance, whose capacitive component is in the 160–320 pF range. The choice of C_1 = 10 pF, C_2 = 150 fF, thus $G_1 \approx -67$, results in an upper limit of 0.5 pF for the parasitic capacitance C_p. which guarantees sufficient margin to keep under control the $1/f$ noise at a cost of a marginal 10% increase in the overall amplifier noise with respect to the ideal OTA [see Equation (1)]. In such a design, $\overline{E^2}_{n_{eq}}$ and $\overline{E^2}_{n_{eq,OTA}}$ are almost the same; therefore, both terms will be denoted with the same symbol, $\overline{E^2}_{n_{eq}}$.

Table 1. Transistor operating points.

Transistor	$V_{GS} - V_T$ [mV]	IC	g_m [μA/V]	r_o [MΩ]
M_p	110	0.075	52.3	5.1
M_n	490	51	7.76	108
M_{cas}	24	0.36	43.7	8.2

To minimize the noise, the first OTA stage has been designed as a simple telescopic cascode amplifier (seeFigure 3). This well-known configuration guarantees excellent noise performance thanks to its few transistors, while cascoding and proper transistor sizing ensure enough gain. Straightforward analysis relates the input-referred noise power spectral density to the noise sources of the transistors, leading to:

$$\overline{E^2}_{n_{eq}} \cong \frac{8kT\gamma}{g_{mp}} + \frac{8kT\gamma g_{mn}}{(g_{mp})^2} + \frac{2K_p^{(1/f)}}{C'_{ox}W_pL_p}\frac{1}{f} + \frac{2K_n^{(1/f)}}{C'_{ox}W_nL_n}\left(\frac{g_{mn}}{g_{mp}}\right)^2\frac{1}{f} \qquad (2)$$

The result has been derived by taking into account that the transistors M_{bias} and M_{cas} do not significantly contribute to the overall noise. To minimize the noise, the condition $g_{mp} \gg g_{mn}$ must be fulfilled, *i.e.*,$(W/L)_p \gg (W/L)_n$. Under this assumption the total noise power density results:

$$\overline{E^2}_{n_{eq}} = \frac{8kT\gamma}{g_{mp}} + \frac{2K_p^{(1/f)}}{C'_{ox}W_pL_p}\frac{1}{f} \qquad (3)$$

with γ equal to 2/3 for transistors working in strong inversion or $1/(2\kappa)$ in weak-inversion ($\kappa \approx 0.7$) [12]. In order to minimize the thermal noise without increasing the current consumption, the M_p transistors have to work in the weak inversion region, where the transconductance is maximum. Its value can be estimated from the EKV model [13], valid in all regions of inversion:

$$g_m \cong \frac{\kappa I_D}{U_T}\frac{2}{1+\sqrt{1+4\,IC}} \qquad (4)$$

U_T and IC being the thermal voltage and the inversion coefficient [12], respectively. The latter parameter is defined as:

$$IC = \frac{\kappa I_D}{2\mu C'_{ox}\left(\frac{W}{L}\right)U_T^2} \qquad (5)$$

and is much less than 1 in weak-inversion or subthreshold region. >From Equation (4), the overall-amplifier input-referred thermal noise power spectral density results:

$$\overline{E^2}_{n_{th}} = \frac{8kTU_T}{\kappa^2 I_{bias}} \qquad (6)$$

with I_{bias} being the bias current of the telescopic cascode amplifier. Assuming a first-order roll-off of the frequency response, the noise bandwidth is $\frac{\pi}{2}\cdot 10$ kHz $\cong 15.7$ kHz. For an upper limit of 3 μV_{rms} for the thermal noise contribution in this noise band, a minimal I_{bias} value of about 3 μA is required; to be conservative, we set I_{bias} = 4 μA .

With regard to the flicker noise, its contribution can be reduced by increasing the PMOS transistor area ($W_p \times L_p$). Considering that for the adopted technology $K_p^{(1/f)} \approx 2\times10^{-26}$ V^2F and $C'_{ox} \approx$ 5 fF/μm^2 and setting a noise corner frequency lower than 100 Hz, the input PMOS transistors were sized with $W_p \times L_p = 400\times1\,\mu m^2$. This sizing results into a stray OTA input capacitance C_p of approximately 500 fF, which does not excessively impair the equivalent input noise. Moreover, in order to fulfill the requirement $(W/L)_p \gg (W/L)_n$, the settings $W_n = 5\,\mu$m and $L_n = 40\,\mu$m were applied, thus forcing these devices to work in the strong inversion region (see Table 1). For this choice of the transistor lengths, cascoding of the input differential pair is needed to not to degrade the output resistance and, thus, to not to lower the amplifier gain. The introduction of M_{cas} guarantees an output resistance $R_{out} = r_{0n}||(g_{mcas}r_{0cas}\,r_{0p}) \approx r_{0n}$ and an amplifier gain $g_{mp}r_{0n} \approx 75$ dB.

In summary, by careful transistor sizing, the noise within the amplifier band may be reduced to the sole thermal noise. Since in a generic amplifier a trade-off exists between noise and current consumption, its efficiency may be evaluated referring to the noise efficiency factor (*NEF*), which is a parameter commonly adopted in literature, proposed in [14] and defined as:

$$NEF = V_{in,rms}\sqrt{\frac{2I_{tot}}{\pi U_T\, 4kT\, BW}} \tag{7}$$

where $V_{in,rms}$ is the input-referred rms noise, I_{tot} is the total supply current and *BW* is the bandwidth of the amplifier. In this limit, the minimum *NEF* achievable with a simple differential stage can be estimated from Equations (6) and (7) leading to a value of $\sqrt{2}/\kappa = 2.02$ [15]. However, in our design the contribution to the thermal noise of the current mirror transistors cannot be completely neglected. In fact, even if M_p transistors work in weak-inversion and M_n FETs are biased in strong inversion region, the transconductance ratio is only about 7 (see Table 1). Considering the general expression of the transconductance given by Equation (4) and the total input-referred thermal noise given by the first two terms of Equation (2), the theoretical *NEF* for this preamplifier becomes:

$$NEF_{th} = \frac{\sqrt{2}}{\kappa} \cdot \sqrt{1 + \frac{2\gamma\kappa}{\sqrt{IC_n}}} \cong 2.15 \tag{8}$$

where $\gamma \approx 2/3$, $\kappa \approx 0.7$ and IC_n is the inversion coefficient of current mirror transistors. In reality, the *NEF* of the proposed amplifier is slightly larger compared to this theoretical limit for three main reasons:

- The transconductance of the input transistors is slightly lower than $\kappa I_D/U_T$ since their inversion coefficient is larger than zero. For the present design, the input transistor *IC* is 1.075 (see Table 1), and consequently the rms input noise and the *NEF* increase by a factor of 1.034.
- The input-referred noise of the overall amplifier is larger by a factor $(1+1/G_1+C_p/C_1)^2$ than the one of the first operational amplifier, as stated by Equation (1). Taking into account that the parasitic input capacitance is approximately 0.5 pF, mainly due to the OTA input transistors, this factor is equal to 1.065. A further contribution derives from the strays associated with the input capacitor plates that was drastically reduced by connecting the capacitor bottom plate (which has the largest parasitism) to the amplifier input and by connecting the top plate to the OTA terminal. In this way, a parasitic capacitance larger than 1.5 pF was avoided.
- The current drawn by the second amplifying stage contributes to the total current in Equation (7) but does not lower the input-referred thermal noise. Therefore, the *NEF* increases by a factor of $\sqrt{(1+I_2/I_1)} \cong 1.037$, where I_1 and I_2 are the currents drawn by the first and the second operational amplifier respectively.

These three factors increase the estimated *NEF* of the overall amplifier to about 2.45.

Finally, note that the adoption of a telescopic cascode amplifier, known to have a small output dynamic, is not a limiting factor for the input signal amplitude. In fact, the positive voltage swing is ≈ 450 mV $(|V_{T,p}|+V_{cas})$ while the negative swing is about 1 V $(1.5\,V - V_{ov,n})$, thus assuring a maximum amplitude for the input signal of about 7 mV.

3.1.2. Second Amplifying Stage

A second gain stage is needed to increase the amplitude level of the input signal before multiplexing. Since this stage requires a relative large output swing, a two-stages OTA can be employed. Provided that the first stage has enough gain, the impact of this second amplifier on the input-referred noise is negligible, and its power dissipation can be reduced without affecting the noise performance. In fact, the noise of the second op-amp is dominated by the input transistors, designed to operate in weak-inversion region. If we limit the contribution to the equivalent input noise added by the second stage OTA to less than 0.3 μV_{rms} (*i.e.*, 1/10 of the dominant contribution), a minimum current of 100 nA is needed in the input differential pair of the second op-amp. To be conservative, we set the bias current to 200 nA in the first stage, while a current of 100 nA is drawn by the second stage.

Finally, nonlinear distortion may be of some concern in the second stage. Distortion arises from the non-linear high-resistance pseudo-resistors placed in the feedback path, which is driven by a large output voltage swing. For this reason the subthreshold MOS transistors were sized to have resistance values an order of magnitude higher than those in the first stage. This choice makes the signal current flowing through them always several orders of magnitudes lower than the signal through the capacitors, limiting the measured Total Harmonic Distortion (THD) generated by the stage below 5% even if the pseudo-resistors were modulated by a 1.5 V peak-to-peak 1 kHz sinusoid.

3.1.3. High-Pass Filter Design and Optimization

Let us now consider the noise due to the MOS-bipolar pseudo-resistors that was neglected in Equation (2). It was experimentally verified that the noise spectral density of the pseudo-resistors complies with the usual equation:

$$\overline{S^2}_R = \frac{4kT}{R} \tag{9}$$

where *R* is the small signal incremental resistance. This noise contribution to the OTA input-referred noise power spectral density follows a $1/f^2$ law and is given by:

$$\overline{E^2}_{n_R} = 2\frac{4kT}{R}\frac{1}{(C_1 2\pi f)^2} = \frac{8kT f_1}{2\pi G_1^2 C_2 f^2} \tag{10}$$

where f_1 is the cut-off frequency of the high-pass filter set by the first amplifier stage. The factor of 2 in Equation (10) accounts for the two pseudo-resistors, one in the feedback path and one connected to the positive input terminal of the operational amplifier. For simplicity, let us now neglect the presence of the selective high-pass filter following the first stage. By integrating Equation (10) from f_1 to the low-pass cut-off frequency of the overall amplifier, f_{LP}, we get:

$$\int_{f_1}^{f_{LP}} \overline{E^2}_{n_R} df \cong \frac{8kT}{2\pi G_1^2 C_2} \tag{11}$$

which is independent of the cut-off frequency f_1. From Equation (11) the contribution of the pseudo-resistors to the OTA input-referred noise is about 2.79 μV_{rms}. This value is not negligible and explains why it is not convenient to implement the selective high-pass filter (*i.e.*, the 300 Hz cut-off frequency high-pass filter) in the first stage, as this solution would degrade the noise performance. The G_M -*C* high-pass filter is instead able to cut off most of the noise from the pseudo-resistors of the first stage, but a careful choice of the filter capacitor is needed to reduce its noise contribution. We will discuss this point in detail. Let us now suppose that $f_1 << f_{HP}$, where f_{HP} is the cut-off frequency of the G_M -*C* high-pass filter. For frequency above f_1, the input-referred power spectral density due to both the pseudo-resistors of the first stage and to the G_M -*C* filter can be written as:

$$\overline{E^2}_{eq(1/f^2)} \cong \frac{8kT}{R}\frac{1}{(C_1 2\pi f)^2} + \overline{S_{nI}^2}\frac{1}{(G_1 C_{HP}\, 2\pi f)^2} \tag{12}$$

where $\overline{S_{nI}^2}$ is the output current noise of the high-pass filter. For simplicity, let us assume that the current noise generated by the G_M cell G_M cell $(G_M = 1/R_{HP})$ can be written as $4kT/R_{HP}$.. Equation (12) becomes:

$$\overline{E^2}_{eq(1/f^2)} = \frac{8kTf_1}{2\pi G_1^2 C_2 f^2} + \frac{4kTf_{HP}}{2\pi G_1^2 C_{HP} f^2} \tag{13}$$

By integrating Equation (13) over the amplifier band, *i.e.*, from f_{HP} to f_{LP}, we obtain:

$$\int_{f_{HP}}^{f_{LP}} \overline{E^2}_{eq(1/f^2)}\, df = \frac{8kT}{2\pi G_1^2 C_2}\frac{f_1}{f_{HP}} + \frac{4kT}{2\pi G_1^2 C_{HP}} \tag{14}$$

The result suggests to reduce the first term by minimizing the f_1/f_{HP} ratio while a large capacitor value C_{HP} is needed to reduce the second term. In practice, the G_M cell is a simple differential stage (see Figure 3) whose bias current can be externally tuned. Because a small bias current ($I \cong 1$ nA) is needed to synthesize a high value resistance, all the transistors work in weak-inversion region and their transconductance is $G_M = g_{m1} \cong \kappa I/U_T$. The current noise of this configuration is therefore:

$$\overline{S_{nI}^2} = 4\frac{2kT}{\kappa}g_{m1} = \frac{8kT}{\kappa R_{HP}} \tag{15}$$

The current noise of the G_M cell is therefore a factor of $2/\kappa$ larger than that of a resistor R_{HP} and this factor has to be added in the second term of Equation (14).

In summary, in order to keep the input-referred noise due to the first stage pseudo-resistors and to the G_M cell in the amplifier band lower than 1 μV_{rms}, f_1 has to be at least a factor of 10 smaller than f_{HP} and C_{HP} must be greater than 2 pF. The two pseudo-resistors in the first stage of the circuit in Figure 3 were designed to have $f_1 < 10$ Hz (with $f_{HP} = 300$ Hz) and $C_{HP} = 7$ pF.

3.1.4. Line Buffer and Multiplexer

A circular shift register, controlled by a 625 kHz clock (ck_{mux} in Figure 1), sequentially enables each amplifier to access the common data lead; each pass gate switches on both the rising and the falling edge of the multiplexer clock, avoiding clock transitions in the middle of the sampling window. As in this topology each amplifier has to drive a long routing line connecting all the pass-gate outputs, a class-AB buffer is needed to boost the line. To save power, each buffer is switched off when the corresponding amplifier is not selected and turned on when the previous amplifier is selected (*i.e.*, one clock edge before). In this way, at any time, only two buffers are powered, draining about 25 μA each from the supply. Further amplification is provided before sampling by a variable-gain amplifier (VGA) added to the amplification chain at the multiplexer output. The gain can be digitally varied between 1 to 8 by means of two auxiliary control bits.

3.2. Analog to Digital Converter

The multiplexed and amplified signal is then converted into a digital form by an 8 bit SAR ADC (see Figure 1). This topology is well-suited for low power and small-area applications since it requires a minimal amount of analog circuitry. To save area, analog multiplexing of input signals, requiring only one AD converter, was employed even though it resulted in a small power penalty with respect to the use of one converter per each amplifier and digital multiplexing of data [16].

The number of the converter bits was chosen taking into account the following considerations:

- The maximum amplitude for an extra cellular action potential is ~ 1 mV [17] while the minimum signal is about 10 μV, the latter being determined by the typical input noise due to neural background activity and electrode impedance [18]. Therefore, the ratio between the Full Scale Range (FSR) and the ADC least significant bit, *i.e.*, the dynamic range of the converter, should be better than 1 mV/10 μV $\approx$ 100 . This results in a converter resolution larger than 6.35 bit.

- The ADC quantization noise has to be kept much lower than the minimum detectable signal, *i.e.*, the rms input noise. This requirements translates into:

$$\frac{FSR}{2^n G\sqrt{12}} < 10\ \mu V \tag{16}$$

where *G* is the amplifier gain and *n* is the number of converter bits. The worst-case condition occurs for the minimum gain of the amplification chain, which is set by the ratio between the FSR and the maximum amplitude of the input signal, A_{max} :

$$G_{min} < \frac{FSR}{A_{max}} \cong \frac{3\ V}{1\ mV} = 3000 \tag{17}$$

In the present design we set $G_{min} \cong$ 2000 and thus:

$$n > log_2[\frac{FSR}{\sqrt{12} \cdot 10\mu V \cdot G_{min}}] \cong 5.4 \tag{18}$$

In practice, taking into account that the effective number of bit (ENOB) is always lower than the nominal figure, the use of an 8 bit ADC will provide a safe margin for the design. A higher resolution is useless and harmful since it would increase the ADC power consumption.

The ADC consists of three main parts: the capacitor array, the comparator and the logic block performing the successive-approximation algorithm. The capacitor array, which acts as sampling capacitance during the sampling interval and as charge-sharing DAC during the conversion period, has been implemented as a binary array. The value of the unit capacitance, C_{LSB}, was chosen considering two requirements:

- The noise in the sampling phase has to be significantly smaller than the quantization noise. Thus, the total capacitance of the array, C_{TOT}, must satisfy the condition:

$$\frac{KT}{C_{TOT}} < \frac{(V_{REF})^2}{12} \tag{19}$$

- which clearly is not a limit factor, resulting in C_{TOT} much larger than about 1 fF.
- The LSB capacitance has to be sufficiently accurate and it has to fulfill the following requirement [19]:

$$2\frac{\Delta C_{LSB}}{C_{LSB}} \cdot \sqrt{2^{N-1}} = 2\frac{K_c}{\sqrt{A_c}} \cdot \sqrt{2^{N-1}} < 0.5 \tag{20}$$

where $K_c = 0.5\% \cdot \mu$m and A_c are the Pelgrom mismatch parameter and the least significant bit (LSB) capacitor area, respectively. ¿From Equation (20) A_c has to be larger than $(4.5\ \mu\text{m})^2$, which implies $C_{LSB} > 19.5$ fF, considering a specific capacitance of 1 fF/μm^2.

Taking into account also the effect of parasitic capacitance that can degrade the achievable accuracy, $C_{LSB} = 80$ fF was employed as unit capacitor, resulting in a total capacitance of $2^N \cdot C_{LSB} = 20.48$ pF.

The comparator was implemented as a classical two-stage topology, as depicted in Figure 4. A 20 dB gain preamplifier is followed by a PMOS latch stage and a couple of inverters are connected at the latch outputs to regenerate the signals. The two-stage architecture was preferred to a dynamic latch in order to avoid kickback noise and metastability [20,21].

The successive approximation logic has been implemented as a synchronous full-custom logic. The A/D converter works with a clock frequency of 20 MHz, providing 1.25 MS/s, *i.e.*, about 64 channels $\times 20$ kS/s per channel. Hence, a conversion period lasts 16 clock cycles (see Figure 5): The sampling phase lasts 7 clock periods to relax the specs of the input signal buffer, while every bit decision takes one clock period, starting from the most significant bit (*MSB*). The last clock cycle is reserved for the end-of-count (EOC) operations. The current consumption of input buffer and A/D converter are 250 and 410 μA, respectively.

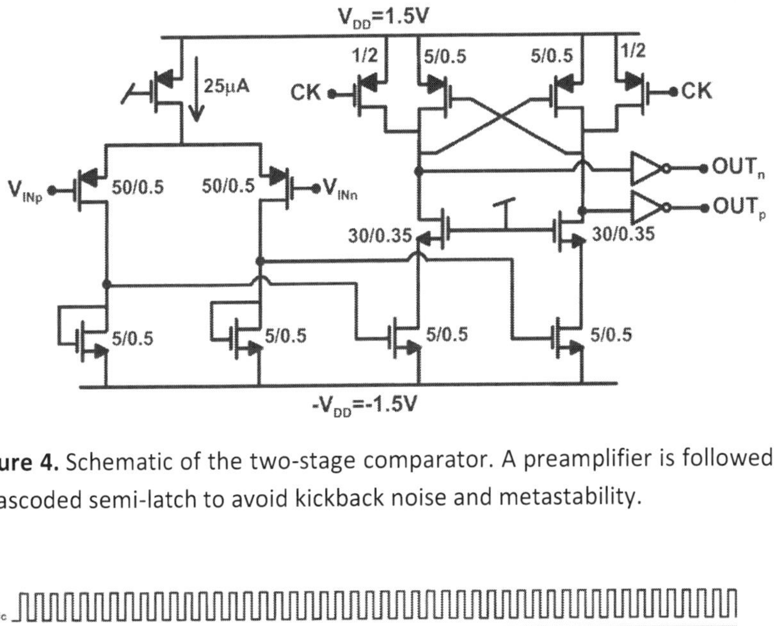

Figure 4. Schematic of the two-stage comparator. A preamplifier is followed by a cascoded semi-latch to avoid kickback noise and metastability.

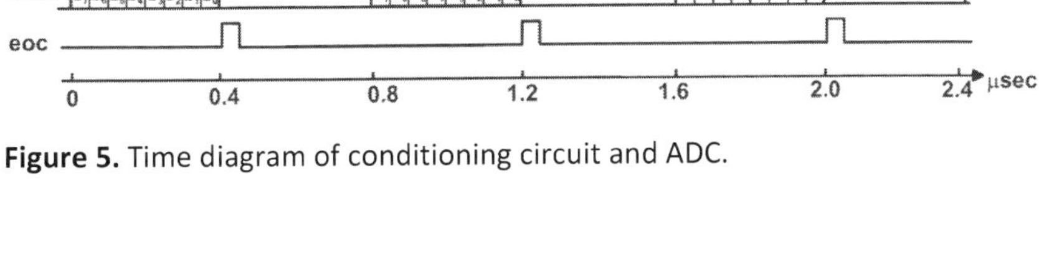

Figure 5. Time diagram of conditioning circuit and ADC.

3.3. Digital Signal Processing

The implemented data reduction system takes advantage of the low duty cycle of the neural activity to eliminate transmission of useless signal containing only noise. Spike duration and firing rates are typically lower than 1 ms and 100 Hz [22], respectively. A finite-state machine (FSM in Figure 6) compares each incoming digital sample with a user-programmable digital threshold, *i.e.*, an 8 bit word stored in the embedded SRAM. When the threshold is crossed, 20 samples of the signal from the same channel are recorded, covering a 1 ms time frame (see Figure 7). This strategy was chosen taking into account that a clear identification of APs requires the detection of three features, namely the amplitude of the peak and the trough and the time interval between them [22], which can be extracted with enough resolution by using 20 samples, even for the fastest spikes. The threshold can be set by the user before the recording starts, choosing among four predefined digital words: "*x*1000000", "*x*0100000", "*x*0010000" and "*x*0001000" (corresponding to 1/2, 1/4, 1/8 and 1/16 of half the ADC full scale range). Since the spike can have a first peak either positive or negative and the baseline is at the middle of the ADC range, the threshold is either positive and negative for each sample. For example, considering the first threshold, the FSM is triggered if the incoming sample is larger than "11000000" or lower than "01000000". If the gain of the amplification chain is set to the maximum value (83 dB $\cong$ 14000), these four thresholds correspond to an input-referred amplitude of ± 54 μV, ± 27 μV, ± 13 μV and ± 7 μV.

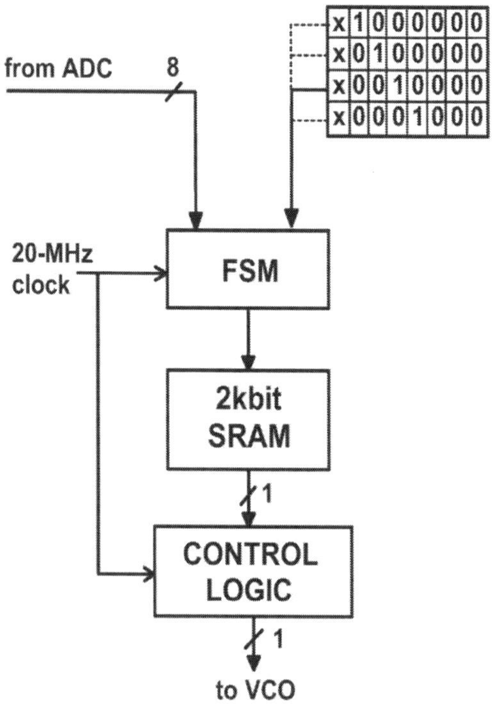

Figure 6. Block diagram of Digital Signal Processing.

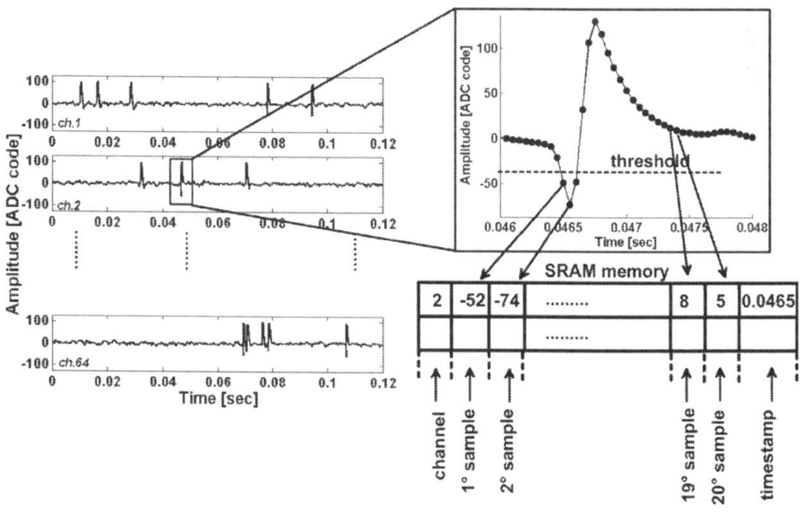

Figure 7. Data compression algorithm: when a spike is detected, 20 digital samples are stored in the embedded memory, together with channel address and timing stamp information.

The recorded samples are stored in a 2 kbit embedded memory (SRAM in Figure 6) together with the channel address and the timing stamp, both with a 8 bit resolution. Each spike is then compressed in 22 bytes. The SRAM is continuously read at 1 Mbit/s rate by the control logic block, which adds service bits for synchronization at the receiver and then performs Manchester encoding of the bit stream. The final data rate reaching the transmitter is 1.25 Mbit/s. The memory read speed was determined by taking into account two aspects:

- the lower the read speed, the higher the compression factor is; 1 Mbit/s read speed results in a compression factor of 10 (from 10.24 Mbit/s, corresponding to 64 channels sampled at 20 kHz per channel and 8 bit per sample, to 1.25 Mbit/s data rate) and allows saving a considerable amount of power and bandwidth at the transmitter side;
- considering the worst-case scenario of 64 channels firing at 100 spike/s and 20 samples per spike with an 8 bit resolution, the average data throughput is about 1 Mbit/s.

The memory was sized to reduce missed spikes when burst activity is present. Monte-Carlo simulations were performed considering that the neural firing rate follows a Poisson statistics. Figure 8 shows the missing spike percentage as a function of the average firing rate for each of the 64 channels and for different memory sizes. Opting for 1 Mbit/s read speed and 2 kbit RAM size, the missing spike percentage is less than 0.1% for 64 channels firing at 50 spikes/s and less than 10% for all channels firing at 100 spikes/s. With these choices, the memory is able to store spikes that occur in the same 1 ms frame in 11 consecutive channels before missing a spike. Each AP is transmitted, once completely stored, in 176 μs and thus the worst-case latency time (time difference between the detection and the transmission of an AP) is about 2.9 ms, which is not an issue both in neuroscience experiments and in future prosthetic applications [23].

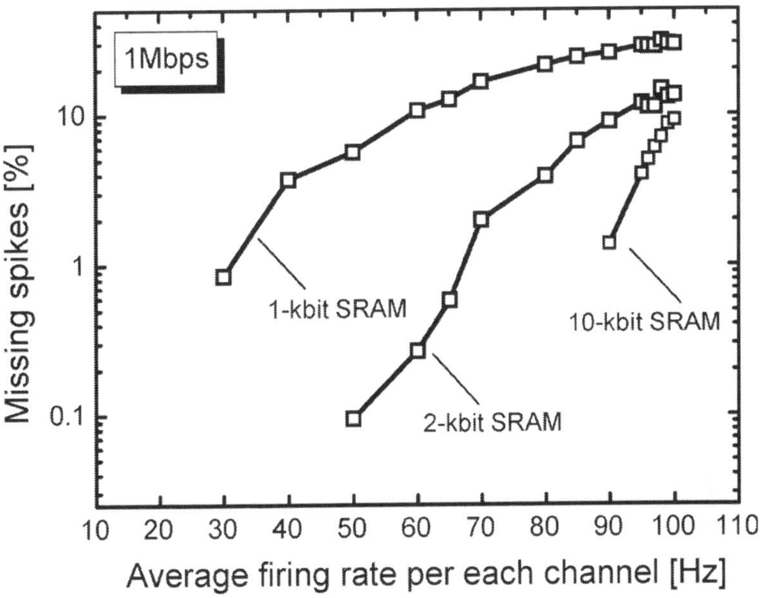

Figure 8. Missing spike percentage as a function of the average firing rate per channel for different memory sizes and for a read speed of 1 Mbit/s.

The validity of the data-reduction algorithm was tested by running a custom clustering software that employed fuzzy c-means spike sorter [24,25] on a data set from a public source [26] with spike waveforms hardly distinguishable one from each other (see Figure 9). The ratio between the peak-to-peak spike amplitude and the rms noise was approximately 10 . Clustering based on Principal Component Analysis (PCA) was performed on original data containing spikes from three different neurons and on the same data but reduced according to the above windowing strategy. Figure 9 shows that clustering on the reduced data set leads to clear separation of the three spike families, with identification errors reported in Table 2. Two types of error occur in a PCA-based clustering: type I errors occur when APs from two different neurons are grouped together (false positives), whilst type II errors occur when not all APs generated by one neuron are grouped together (false negatives). When applied to the original dataset, the PCA-based identification procedure leads to type I/type II mean error rates of 4.1%–5.2%, respectively. The reduced data sets showed 4.6%–5.3% values, thus demonstrating that the reduction strategy preserves the quality needed for effective neuron identification. These results are not surprising considering that, as stated in [22], more than 80% of neural spike information is carried by three analog features of the AP waveform, namely the peak and the through amplitudes and the time between them. The results of spike sorting using these analog features [27] and performed on the same trace are also reported in Table 2. Note that the compression method employed in this work always outperforms the clustering obtained using only these three features. Finally, it is worth nothing that the current consumption of the overall digital signal processing is limited to about 400 μA, most of which consumed by the finite-state machine.

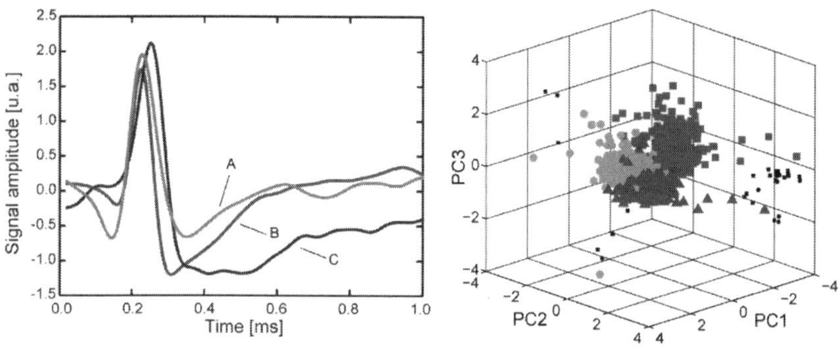

Figure 9. Waveforms of neural spikes and PCA-based clustering results employing the proposed windowing strategy, *i.e.*, processing 20 samples after the threshold crossing.

Table 2. Sorting results.

AP class	Original		Reduced		Peak-trough-width	
Error type	I	II	I	II	I	II
A (379)	4.1%	6.9%	3.1%	2.9%	23%	22%
B (389)	2.1%	3.3%	3.7%	6.9%	12%	18%
C (401)	6.2%	5.5%	6.9%	6.0%	17%	30%

3.4. Manchester-Coded FSK Modulator

The 1.25 Mbit/s bit stream is transmitted using a Manchester-coded binary Frequency Shift Keying (MC-2FSK) modulation at a carrier frequency of 400 MHz. The transmitter (see Figure 10) consists of a voltage-controlled oscillator (VCO) directly modulated by the digital data, inserted in a Phased-Locked Loop

(PLL). Since the modulator is a power-hungry system, the design of the whole PLL, and in particular of the high-frequency oscillator, has to be carried out with great care.

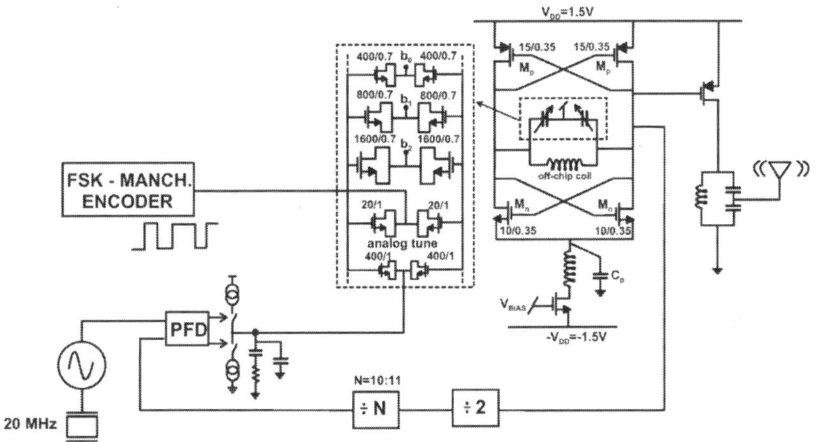

Figure 10. Schematic of the Phased-Locked Loop adopted as frequency synthesizer and modulator. The Manchester-coded data stream directly modulates the VCO. The Power Amplifier adopts an external resonant filter to efficiently deliver power to the antenna allowing long transmission range.

The oscillator has a current-biased topology with double cross-coupled pair transconductor. This oscillator architecture allows to minimize the current consumption for a given phase noise with respect to the topology featuring a single differential pair transconductor [28]. In order to accurately set the PLL output frequency at the desired value of 400 MHz and to compensate for process, temperature and power-supply variations, the oscillator has to span over ±10% of its central frequency, *i.e.*, a band of about ±40 MHz. This means that the capacitance has to be varied as:

$$\frac{|\Delta C|}{C} = 2\frac{\Delta f_0}{f_0} = 40\% \tag{21}$$

f_0 is the central oscillation frequency (400 MHz), Δf_0 the covered band of 80 MHz and C the tank capacitor. To make the oscillation frequency insensitive to the parasitic capacitance of the active devices, routing, pads and package pins (the inductor is off-chip), the tank capacitance has to be set in the pF range. Thus, we designed the oscillator with *C* tunable from 6 pF to 11 pF, determining an inductance value of about 20 nH. The tank capacitance is implemented by three digitally-controlled binary-weighted NMOS inversion-mode varactors that perform coarse tuning of the oscillation frequency and an analog NMOS inversion-mode varactor for fine-tuning of the frequency, enabling to close the loop of the PLL. Another small NMOS varactor is also adopted for frequency modulation.

Once the tank parameters have been determined, the current I_{BIAS} of the tail generator and the (*W/L*) ratio of the pair transistors have to be chosen taking into account the following considerations:

- The small-signal loop gain or excess gain [29] of the oscillator ($g_m R_{TANK}$, where g_m is the small-signal transconductance of the double differential-pair and R_{TANK} is the tank equivalent parallel resistance) has to be larger than 1 in order to assure the oscillation start-up. Since the common-mode output voltage of the oscillator is set to the half of the power supply to maximize the oscillation swing, *i.e.*, $V_{DD}/2 = 1.5$ V, we have:

$$g_m R_{TANK} = \frac{I_{BIAS} R_{TANK}}{V_{ov}} > 1 \tag{22}$$

corresponding to a current larger than $\cong 220$ μA. Note that in Equation (22) it was assumed that the overall transconductance g_m of the double differential-pair is equal to the single MOS transconductance and that NMOS and PMOS transistors have the same overdrive voltage, $V_{ov} = V_{GS} - V_T = (1.5 - 0.7)\text{ V} \cong 0.8\text{ V}$.

The differential oscillation amplitude, A_0, has to be sufficiently large ($>$ 1.5 V) to drive the subsequent stages, *i.e.*, the frequency dividers and the power amplify. Since

$$A_0 \cong \frac{4}{\pi} I_{BIAS} R_{TANK} = \frac{4}{\pi} I_{BIAS} \cdot \omega_0 L Q \tag{23}$$

where *Q* is the tank quality factor [30], the current I_{BIAS} is minimized adopting an inductor with the maximum quality factor. We opted for an off-chip inductor of 21 nH featuring a Q of about 70 at 400 MHz. This choice results in a current larger than 320 μA to achieve the desired voltage amplitude.

- The VCO phase noise, which mainly determines the phase noise of the whole synthesizer at a frequency offset larger than the PLL loop bandwidth, has to be sufficiently lower in order not to worsen the Signal-to-Noise-Ratio (*SNR*) of the modulated signal. This can be evaluated as Single-Sideband to Carrier Ratio, *i.e.*, as the ratio of the power in a 1 Hz bandwidth at an offset ω_m from the fundamental angular frequency ω_0 and the power of the carrier, giving [31]:

$$SSCR(\omega_m) = \frac{kT}{C}\frac{\omega_0}{Q}\frac{1}{\omega_m^2}\frac{1+F}{A_0^2} \tag{24}$$

where F is the noise factor of the transconductor ($F \cong 1$). From behavioral simulations, the phase noise at 1 MHz offset from the carrier has to be lower than –120 dBc/Hz to not to degrade the frequency modulated signal.

- A resonant filter (at twice the oscillation frequency) connected to the source of the NMOS differential pair transistors (see Figure 10 and Figure 11) can be added to reduce flicker noise up-conversion into phase noise, which can severely limit the VCO performance [29].

For $I_{BIAS} = 400$ μA, C in the 6–11 pF range and $Q \cong 70$, the excess gain is approximately 2 and the oscillation amplitude is larger than 2 V, preventing the tail current generator to enter the ohmic region and avoiding a phase noise degradation. Moreover, these choices ensure that the phase noise is not an issue being better than 1–50 dBc/Hz at 1 MHz frequency offset.

The VCO is followed by a cascade of a divide-by-2 dynamic logic prescaler and a static modulo-10 CMOS counter, which further divides the frequency by 10. The PLL reference is the same 20 MHz master clock, allowing the loop to set the oscillation frequency at 20 times the reference clock, *i.e.*, 400 MHz. The PLL also includes a phase-frequency detector with a charge-pump driving an *RC*-*C* off-chip filter (see Figure 10), which sets the loop bandwidth to 50 kHz. Excluding the VCO, the power consumption is mainly due to the high-frequency prescaler, which draws more than 200 μA. The current consumption of the whole synthesizer is about 700 μA.

The synthesizer frequency is modulated in an open-loop mode by the data at 1.25 MHz, *i.e.*, at 1.25 Mbit/s rate with Manchester encoding. Thus, the loop is open for the frequency modulation since the PLL bandwidth is significantly lower than the bit-rate. This prevents the PLL to filter out the induced frequency modulation. The only purpose of the loop is to fix the central oscillation frequency in order to avoid frequency drift due to temperature and power supply variations. To limit the occupied bandwidth, a frequency deviation of about ±400 kHz was adopted. Considering a modulation frequency of 1.25 MHz, this value corresponds to a modulation index of 0.64. The bandwidth of the transmitted signal is determined by the Carson rule [32]:

$$BW = 2 \cdot (\Delta f + f_m) \cong 3.3\ \mathrm{MHz} \tag{25}$$

where Δf and f_m are the peak frequency deviation (400 kHz) and the frequency modulation (1.25 MHz), respectively.

3.5. Power Amplifier

The transmitter is completed by a class-AB power amplifier (PA) that drives a 50- Ω antenna via an off-chip resonant filter. The PA consists of an open-drain PMOS transistor directly connected to the oscillator output, sized in order to draw about 3.5 mA. Thus, the power delivered to the antenna is about 0 dBm considering a reasonable efficiency of 10%. The output power, and thus the power consumption of the PA, were set after estimating the sensitivity of the receiver, which depends on the modulation type, the signal bandwidth and the noise factor of the receiver itself. The receiver input noise in the signal bandwidth of 3 MHz is:

$$N_{in} = kTF \cdot BW = -99\ \mathrm{dBm} \tag{26}$$

where *F* is the noise factor of the receiver, which was assumed to be equal to 10 dB in the above equation. For a low modulation index Manchester-encoded modulation, the *SNR* required at the receiver input to ensure a bit error rate (*BER*) lower than 10^{-5} is $SNR_{min} \cong 20$ dB [33]. Thus the input power needed at the receiver is:

$$P_{min} = N_{in} + SNR_{min} = -79 \text{ dBm} \tag{27}$$

Once the receiver sensitivity has been computed, it is possible to set the PA power to achieve a desired transmission range. Taking into account the Friis transmission equation, the ratio of power available at the input of the receiving antenna, P_R, to the output power of the transmitting antenna, P_T, is:

$$\frac{P_R}{P_T} = G_T G_R \left(\frac{\lambda}{4\pi d}\right)^2 \tag{28}$$

Figure 11. Photo of the board adopted during the neuroscience experiments with rats and microphoto of the die.

where G_T and G_R are the gains with respect to an isotropic radiator of the transmitting and receiving antennas, respectively, λ is the wavelength, and d is the distance between the antennas. While the receiver antenna is a dipole antenna and its gain can be estimated to be 0 dB, the transmitter antenna is a whip antenna with a gain that can be as low as –20 dB. Note that Equation (28) represents the power at the receiver antenna that is supposed to be matched with the receiver. Thus, the receiver input power is 3 dB lower. Thus, considering Equation (28) and setting the maximum transmission distance to 10 m, the power needed at the transmitter antenna is 11 dBm, where a 3 dB factor accounting for the matching between the receiver antenna and the receiver itself is considered. Finally, to have a safety margin, a PA featuring a 0 dBm output power was designed.

4. EXPERIMENTAL RESULTS

4.1. Electrical Characterization

The chip was fabricated in AMS 0.35 μm, 3 V 4M2P CMOS process and occupies a total area of 3.1×2.7 mm^2, pads included (see Figure 11). The preamplifier has a mid-band gain of about 65 dB, a high frequency cut-off of 10.5 kHz and a low-frequency cut-off tunable from 1 Hz to 1 kHz (see Figure 12). The input-referred noise, for $f_{HP}=300$ Hz, is 3.05 μV, while the current consumption is 4 μA for the first stage, 0.001 μA and 0.3 μA for the G_M-C high-pass filter and the second amplifying stage respectively, resulting in an overall *NEF* of 2.5. The input-referred power spectral density is shown in Figure 13 for different values of the G_M-C filter high-pass corner frequency. The plateau of –152 dBV^2/Hz corresponds to an input-referred noise of about (25 nV)^2/Hz, close to the expected value. Note that, for $f_{HP}=300$ Hz, the $1/f^2$ input-referred noise is mainly due to the G_M-C high-pass filter, as predicted by Equation (14). The common-mode rejection ratio (*CMRR*) and the power supply rejection ratio (*PSRR*) of the overall pre-amplifier are higher than 65 dB and 50 dB, respectively, while the cross-talk between adjacent channels due to the analog multiplexing is lower than –40 dB.

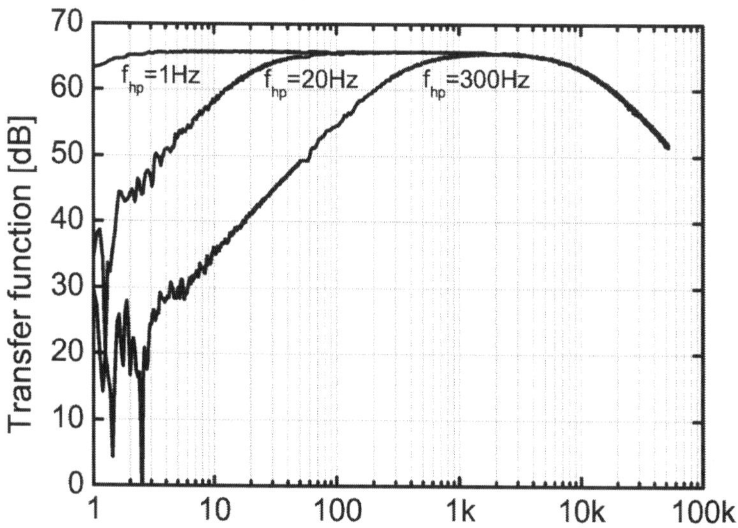

Figure 12. Measured transfer function for different high-pass corner frequencies set by the G_M -C filter.

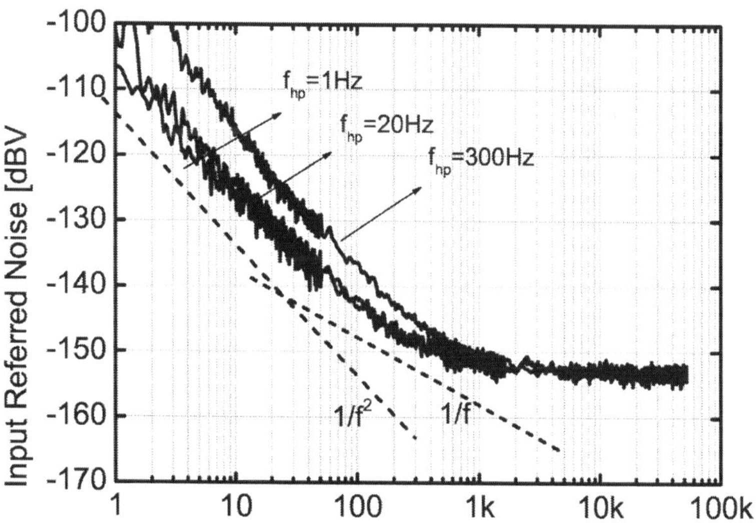

Figure 13. Measured input-referred noise for different high pass-corner frequency.

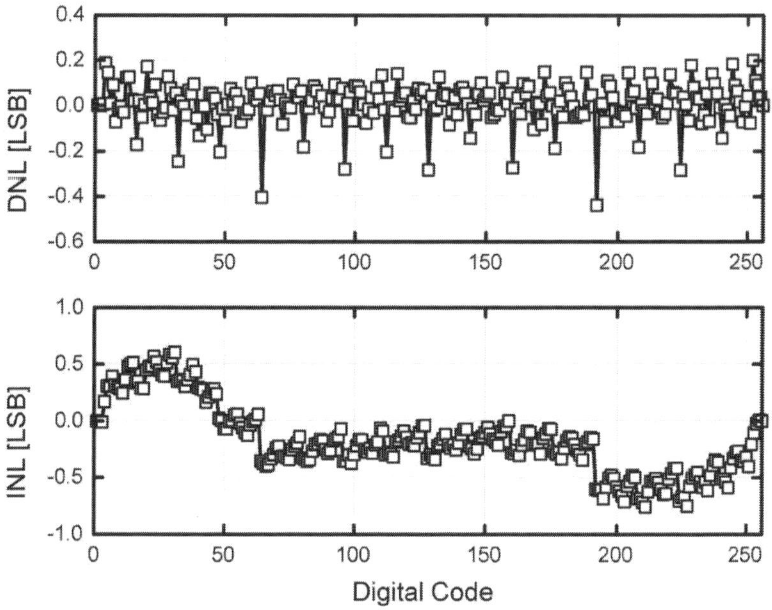

Figure 14. Measured DNL and INL of the integrated AD converter.

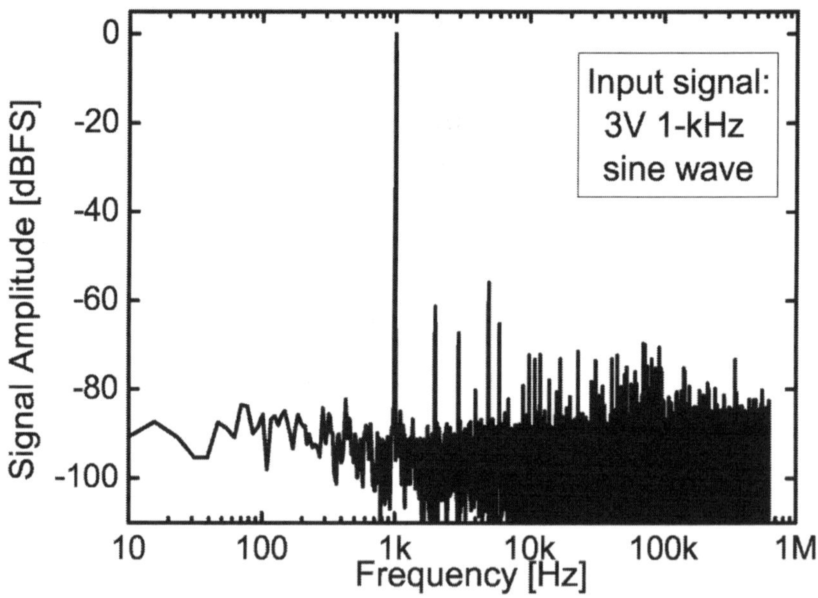

Figure 15. Spectrum of the ADC output for a 3 V sine wave at 1 kHz frequency as input signal.

With regard to the AD converter, Figure 14 shows the results of static measurements. The converter features a DNL less than 0.45 LSB and an INL less than 1 LSB. Figure 15 shows the output spectrum of the ADC for a full-scale sine wave at 1 kHz frequency as input signal. The signal to noise plus distortion ratio (SNDR) is close to 45 dB, and the number of effective bit (ENOB) is thus about 7.2.

Considering the modulator, the frequency range covered by the VCO spans from 310 to 460 MHz (see Figure 16), allowing to lock the PLL loop to a frequency of 400 MHz frequency even with a large spread of tank component values and for sizable variations in power supply and temperature. When no data modulation is performed, the modulator phase noise (see Figure 17) is less than −123 dBc/Hz at 1 MHz offset from the carrier. The phase noise spectrum has a typical low-pass shape, with 50 kHz cut-off frequency, since it is completely dominated by the charge-pump noise [34], while the VCO phase noise is much lower. The rms integral frequency noise of the modulator in the 10 Hz to 10 kHz band is less than 3 kHz (see the residual frequency FM in Figure 17), whereas the output spectrum of the synthesizer when acting as frequency modulator is depicted in Figure 18: more than 98% of the signal power is included in a 3 MHz band.

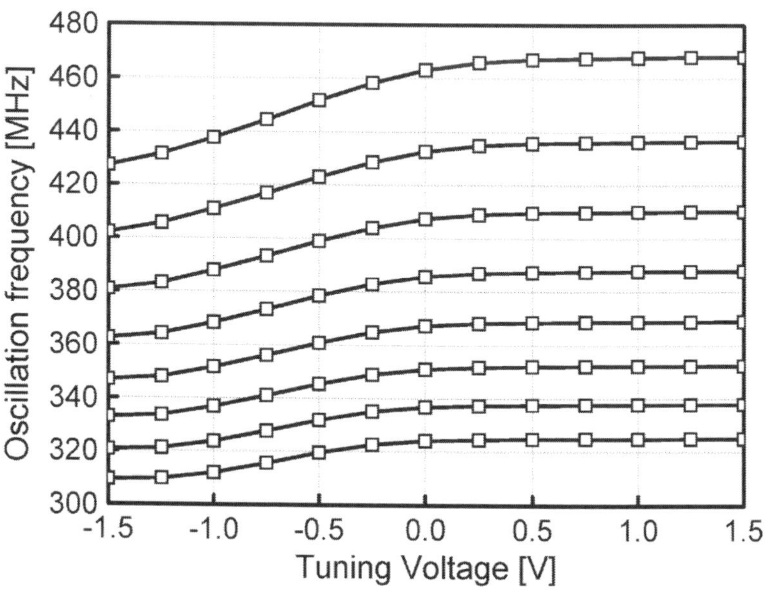

Figure 16. Oscillation frequency varying the analog tuning voltage and for different coarse tuning bit configuration.

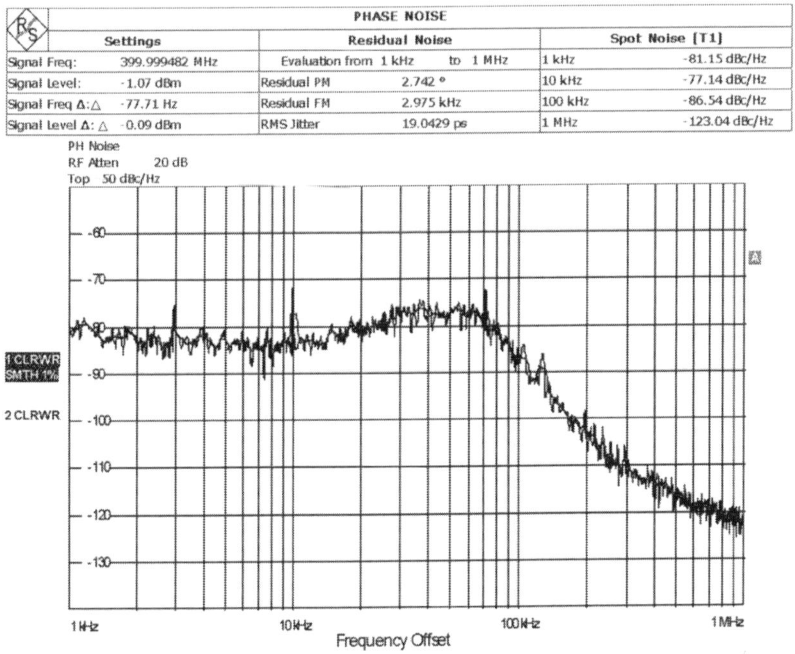

Figure 17. Measured phase noise of the phased-locked loop when the modulation is off.

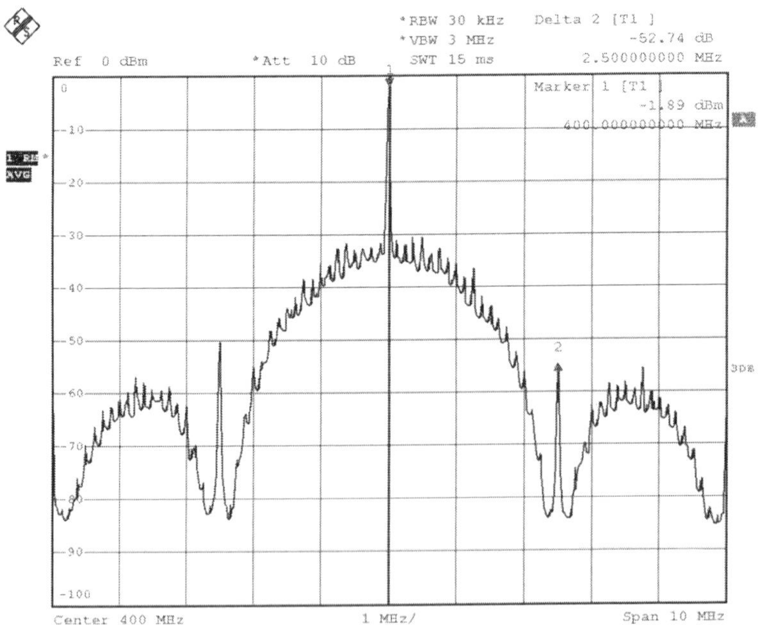

Figure 18. Spectrum of the modulated signal.

For a $\lambda/4$ whip antenna at the transmitter and a dipole antenna at the receiver side, the received power in a free space is shown in Figure 19. The sensitivity of the receiver has been measured to be –74 dBm for a *BER* of 10^{-5}, close to the estimated theoretical value predicted by Equation (27) of –79 dBm, allowing a transmission range larger than 30 m. The system has also been successfully tested for a 10 m distance between transmitter and receiver in a hostile environment as a neuroscience laboratory, giving a PA efficiency of about 7% while delivering a –2 dBm at a 50- Ω output load. Disabling the PA, the tank inductor may act as transmitting antenna; in such measurements about –70 dBm was received by a loop antenna placed at approximately 5 cm from the tank inductor.

The overall performance of the implemented system is provided in Table 3. The power consumption of the presented circuit has been computed by projecting the performance to a complete 64 channel system. To this end, the additional budget needed for the missing 48 amplifiers has been added to the actual power consumption (16.6 mW) leading to 17.2 mW (6.7 mW with the PA turned off). The estimate is fair since all the other circuit blocks work with the same data rate of a full 64 channel system.

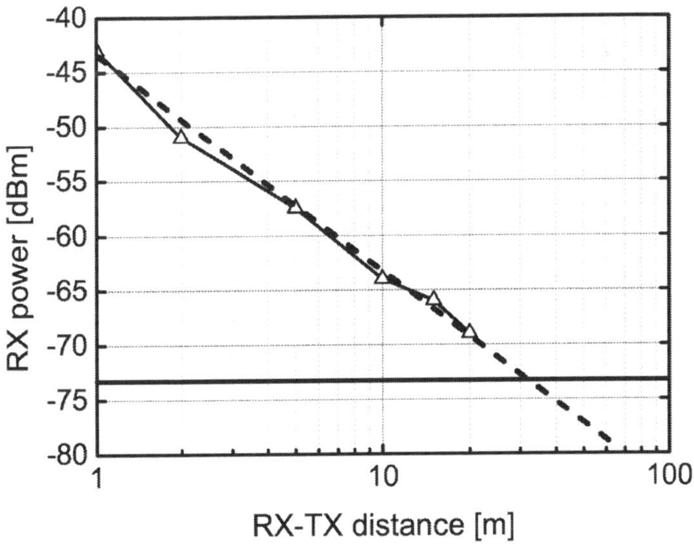

Figure 19. Received power as a function of the distance between transmitter and receiver antenna in free-space. A dipole antenna was adopted at the receiver side.

Table 3. Performance summary.

Technology	0.35 μm AMS
Supply Voltage	3 V
Number of channels	64 (16 available)
Pre-amplifier gain	65 dB
Pre-amplifier band	HP tunable-10.5 kHz
Input-Referred Noise	3.05 μV_{rms} ($f_{HP} = 300$ Hz)
ADC DNL-INL	< 0.45 LSB; < 1 LSB
ADC ENOB	7.2
Transmission frequency	400 MHz
Bit-rate	1.25 Mbit/s
Current Consumption	
20-MHz crystal oscillator	90 μA
Pre-amplifiers	4.3 μA $\times$ 64 $=$ 275 μA
Line buffers	25 μA $\times$ 2 $=$ 50 μA
VGA	50 μA
ADC buffer	250 μA
ADC	410 μA
DSP+2 kbit RAM	400 μA
Modulator (PLL)	700 μA
Power amplifier	3.5 mA
Total power consumption	6.7 mW (17.2 mW with PA enabled)

4.2. In Vivo Experiments

Figure 20 shows the system mounted on a rat for *in vivo* signal acquisition. It consists of two parts, a wireless neural recording headstage and a backpack. To limit disturbances collected from the environment, a small board (3×3 cm^2 and 4.5 g of weight) is directly connected to a 16 channel microelectrode array (Tucker-Davis), with an impedance in the 20–60 k Ω range at 1 kHz, implanted into the somatosensory cortex of an adult rat. The board includes the packaged chip and 10 external small surface-mount components (0805 or 0603 footprint): the VCO inductor, the power-amplifier resonant filter components (2 capacitors and 1 inductor), the 20 MHz quartz, the PLL loop-filter (1 resistor and 2 capacitors) and 2 decoupling capacitors between power supplies. The antenna is a quarter-wavelength whip antenna: it consists of a piece of wire about 17 cm long, easily placed along the back of the rat. The backpack has a weight of 40 g and includes two AAA batteries with 1000 mA/h capacity that may allow neural activity recordings for more than 100 hours. The two batteries are connected via three 10 cm long wires to the headstage. At the beginning of the *in vivo* experiment, the quality of the signals from the implanted electrodes was analyzed using a commercial acquisition system by recording the raw signals and noise from the same 16 electrode channels. The recorded traces featured a noise of about 10 μV_{rms} on all channels. The threshold of the digital peak processor was set to "*x*1000000" and the gain of the overall amplifying chain to the maximum value (83 dB) since the peak-to-peak spike amplitude was always lower than 100 μV. Thus, the digital threshold corresponds to an input-referred amplitude of about ±30 μV, *i.e.*, ±3 times the rms noise. The high-pass cut-off frequency of the front-end amplifiers was set to about 300 Hz to properly reject the low-frequency signals and the power-line noise, enabling correct spike detection. The receiver dipole antenna has been placed at two meters from the rat while neural activity was recorded. Figure 21 shows a single trace recorded during the experiment and two spike waveforms extracted from the data stream, while in Figure 22 a large number of spikes recorded on the same channel are shown aligned in time.

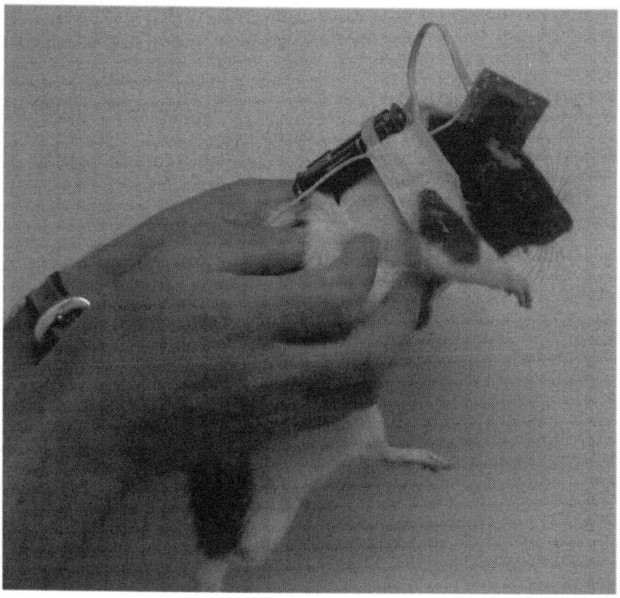

Figure 20. Arrangement for *in vivo* measurements. Note the headstage, the battery backpack and the whip antenna placed along the back of the animal.

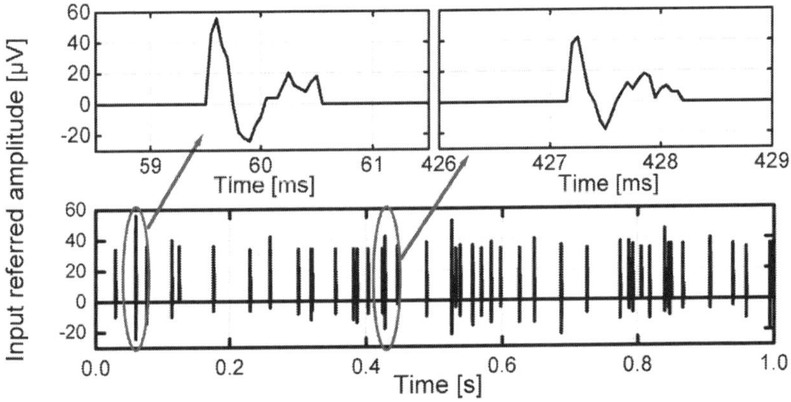

Figure 21. Example of a reconstructed trace registered in an *in vivo*experiment.

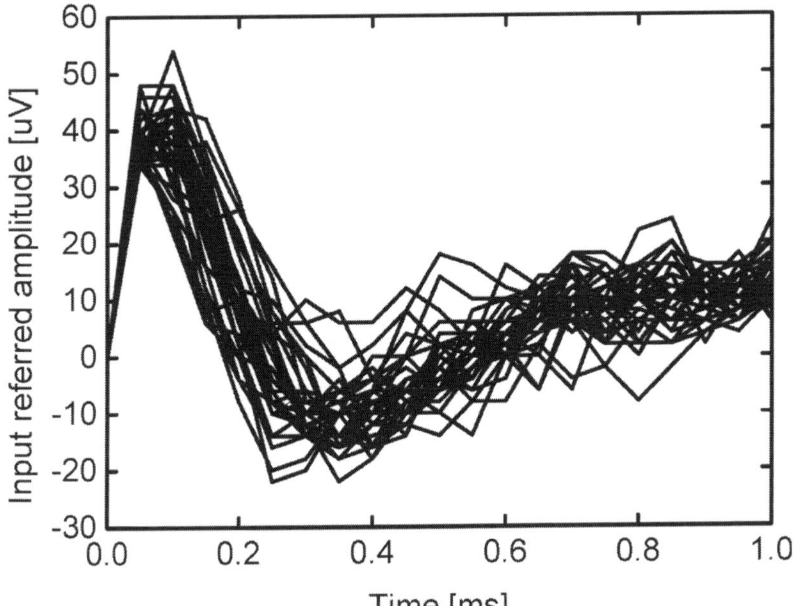

Figure 22. Action potentials registered on a same channel and aligned in time.

5. DISCUSSION AND CONCLUSIONS

Table 4 compares the implemented system to other wireless neural recording ICs with respect to the three most important features, *i.e.*, the transmitted data type (raw data, spike detection or spike waveform), the transmission range and the power consumption per channel. For a fair comparison, since state-of-the-art wireless neural recording systems feature a transmission range always lower than 1 m, the proposed circuit can be thought sized to get a similar transmission range. Since the power needed at the antenna is inversely proportional to the square of the transmission distance, the PA power consumption could be scaled down by a factor of 900, *i.e.*, (30 m/1 m)^2. In this case the equivalent power consumption of the whole system reduces to about 6.7 mW, corresponding to less than 105 μW per channel.

Table 4. Comparison of wireless neural recording ICs.

Parameter	[3]	[6]	[7]	[35]	This work
Technology	0.5 μm	0.35 μm	0.5 μm	0.13 μm	0.35 μm
Power source	inductive link	battery	battery	NA	battery
Number of channels	100	128	32	64	64 (16 avail.)
Overall gain	60 dB	57–60 dB	68–78 dB	54–60 dB	65–83 dB
Input noise	5.1 μV	4.9 μV	9.3 μV	6.5 μV	3.05 μV
TX frequency	433 MHz	4 GHz	915 MHz	915 MHz	400 MHz
Modulation	2FSK	IR-UWB	2FSK (PWM)	FSK-OOK	2MC-FSK
Data type	spike detection	raw data	raw data	raw data	AP waveform
Data rate	330 kbit/s	90 Mbit/s	640 kbaud/s	1.5 Mbit/s	1.25 Mbit/s
Bandwidth	0.8 MHz	1 GHz	38 MHz	3 MHz	3 MHz
TX range	13 cm	¡1 m	1m	1 m	30 m
Power per channel	135 μW/ch	47 μW/ch	220 μW/ch	80 μW/ch	269 μW/ch

By comparing our system with the one described in [3], working in the same frequency range, the proposed system features a lower power consumption per channel. Also the data quality is greatly improved since spike detection is replaced by waveform detection, making possible single spike identification over a large microelectrode array. Compared with the IC in [7], our system shows an improvement of a factor larger than 2 in terms of power consumption per channel. It is true that the data rate does not allow all raw data acquisition as in [7], but our design optimizes information transfer and is potentially compliant with the narrowband spectral limitations in the Medical Implant Communication Service (*MICS*) band range (402–405 MHz). In the future, this system could also be designed to work in the Industrial Scientific and Medical (*ISM*) band (902–928 MHz).

The system in [6] is better in term of power per channel but, relying on an UWB transmitter, it can suffer from interference. Moreover, this system has never been tested in *in vivo* experiments and no information on the adopted antenna or on the power delivered by the PA is given. Regarding the system presented in [35], its power consumption per channel (80 μW/ch) was evaluated considering the minimum transmitter power, assuming that it allows a 1 m transmission distance. However, the data rate is limited to 1.5 Mbit/s, translating into data acquisition at a sampling rate $<$ 3 kS/s from all 64 channels. Higher sampling rates are possible only for a limited number of channels: assuming that 20 kS/s is the minimum sampling rate to acquire good quality neural signals, the maximum number of channels that can be recorded and transmitted simultaneously is only 9.

In conclusion, thanks to the implemented compression algorithm that allows to preserve the information needed for single neuron identification while limiting the data throughput and the bandwidth, we achieved a good balance between power consumption and quality of transmitted data and transmission range. In addition, the power consumption per channel was lowered by a careful sizing of each individual circuit block.

ACKNOWLEDGMENT

The authors would like to thank L. Fadiga (IIT, Genova) for useful discussions and A. Oleynjk (Universita' di Ferrara) for providing us the clustering software.

REFERENCES AND NOTES

1. Olsson, R.H.; Buhl, D.L.; Sirota, A.M.; Buzsaki, G.; Wise, K.D. Brain-machine interfaces: Past, present and future. *Trends Neurosci.* **2006**, *29*, 536–546.
2. Seese, T.; Harasaki, H.; Saidel, G.; Davies, C. Characterization of tissue morphology, angiogenesi, and temperature in the adaptive response of muscle tissue to chronic heating. *Labs Invest.* **1998**, *78*, 1553–1562.
3. Harrison, R.R.; Watkins, P.T.; Lovejoy, R.O.; Kier, R.J.; Black, D.J.; Greger, B.; Solzbacher, F. A low-power integrated circuit for a wireless 100-electrode neural recording system. *IEEE J. Solid State Circuits* **2007**, *1*, 123–133.
4. Sodagar, A.M.; Perlin, G.E.; Yao, Y.; Najafi, K.; Wise, K.D. An implantable 64-channel wireless microsystem for single-unit recording. *IEEE J. Solid State Circuits* **2009**, *9*, 2591–2604.
5. Hochberg, L.R.; Serruya, M.D.; Friehs, G.M.; Mukand, J.A.; Saleh, M.; Caplan, A.H.; Branner, A.; Chen, D.; Penn, R.D.; Donoghue, J.P. Neuronal ensemble control of prosthetic devices by a human with tetraplegia.*Nature* **2006**, *442*, 164–171.

6. Chae, M.S.; Liu, W.; Yang, Z.; Chen, T.; Kim, J.; Sivaprakasam, M.; Yuce, M.A. 128-Channel 6 mW Wireless Neural Recording IC with On-the-Fly Spike Sorting and UWB Transmitter. In Proceedings of the IEEE International Solid-State Circuits Conference (ISSCC 2008)Digest of Technical Papers, San Francisco, CA, USA, 3–7 February 2008; pp. 146–148.
7. Lee, S.B.; Lee, H.M.; Kiani, M.; Jow, U.M.; Ghovanloo, M. An inductively powered scalable 32-channel wireless neural recording system-on-a-chip for neuroscience applications. *IEEE Trans. Biomed. Circuits Syst.* **2010**, *4*, 360–371.
8. Liu, Y.-H.; Li, C.-H.; Lin, T.-H. A 200-pJ/b MUX-Based RF transmitter for implantable multichannel neural recording. *IEEE Trans. Microw. Theory Tech.* **2009**, *57*, 2533–2541.
9. Bonfanti, A.; Ceravolo, M.; Zambra, G.; Gusmeroli, R.; Borghi, T.; Spinelli, A.S.; Lacaita, A.L. A multi-channel low-power IC for neural spike recording with data compression and narrowband 400-MHz MC-FSK wireless transmission. In Proceedings of the 2010 European Solid-State Device Research Conference (ESSCIRC), Seville, Spain, 14–16 September 2010; pp. 330–333.
10. Harrison, R.R.; Charles, C. An implantable 64-channel wireless microsystem for single-unit recording. *IEEE J. Solid State Circuits* **2003**, *6*, 958–965.
11. C_p is not drawn in Figure 3. It can be easily verified that these strays do not affect, in principle, the differential gain of the first stage. Only their mismatch causes a pole-zero doublet which, however, has no practical impact since it lays at very low frequency, below the band of interest.
12. Vittoz, E.; Fellrath, J. CMOS analog integrated circuits based on weak inversion operations. *IEEE J. Solid State Circuits* **2003**, *3*, 224–231.
13. Enz, C.C.; Krummenacher, F.; Vittoz, E.A. An analytical MOS transistor model valid in all region of operation and dedicated to low-voltage and low-current applications. *Analog Integr. Circuits Signal Process.* **1995**, *8*, 83–114.
14. Steyaert, M.S.J.; Sansen, W.M.C. A micropower low-noise monolithic instrumentation amplifier for medical purposes. *IEEE J. Solid State Circuits* **1987**, *22*, 1163–1168.

15. This limit can be overwhelmed adopting a current-reuse input stage that makes use of both NMOS and PMOS transistors [6,36]. However, in this case, the impact of flicker noise and of the input capacitance [see Equation (1)] may be detrimental.

16. Chae, M.S.; Liu, W.; Sivaprakasam, M. Design Optimization for Integrated Neural Recording Systems. *IEEE J. Solid State Circuits* **2008**, *49*, 1931–1939.
17. Mollazadeh, M.; Murari, K.; Cauwenberghs, G.; Thakor, N. From Spikes to EEG: Integrated Multichannel and Selective Acquisition of Neuropotentials. In Proceedings of the 30th Annual International Conference of the IEEE Engineering in Medicine and Biology Society (EMBS 2008), Vancouver, Canada, 20–25 August 2008; pp. 2741–2744.
18. Joye, N.; Schmid, A.; Leblebici, Y. An electrical model of the cell-electrode interface for high-density microelectrode arrays. In Proceedings of the 30th Annual International Conference of the IEEE Engineering in Medicine and Biology Society (EMBS 2008), Vancouver, Canada, 20–25 August 2008; pp. 559–562.
19. Agnes, A. Very low power Successive Approximation A/D Converter for Instrumentation Applications. Ph.D. Dissertation, Universitá di Pavia, Pavia, Italy, 2009.
20. Razavi, B.; Wooley, B.A.; Leblebici, Y. Design techniques for high-speed, high-resolution comparators. *IEEE J. Solid State Circuits* **1992**, *27*, 1916–1926.
21. Figueiredo, P.M.; Vital, J.C. Kickback noise reduction techniques for CMOS latched comparators. *IEEE Trans. Circuits Syst. II Express Briefs* **2006**, *53*, 541–545.
22. Vibert, J.F.; Costa, J. Spike separation in multiunit records: a multivariate analysis of spike descriptive parameters. *Electroenceophal. Clin. Neurophys.* **1979**, *47*, 172–182.
23. Oweiss, K. A Systems Approach for Data Compression and Latency Reduction in Cortically Controlled Brain Machine Interfaces. *IEEE Trans. Biomed. Eng.* **2006**, *53*, 1364–1377.
24. Lewicki, M.S. A review of methods for spike sorting: The detection and classification of neural action potentials. *Comput. Meth. Prog. Biomed.* **1998**, *9*, 53–78.
25. Zouridakis, G.; Tam, D.C. Identification of reliable spike templates in multi-unit extracellular recordings using fuzzy clustering. *Comput. Meth. Prog. Biomed.* **2000**, *61*, 91–98.

26. The public database is available at Professor Rodrigo Quiroga's website, http://www2.le.ac.uk/departments/engineering/research/bioengineering/neuroengineering-lab/spike-sorting/ [37]. The file adopted to validate the compression algorithm is *C-difficult1-noise01.mat*.
27. Bonfanti, A.; Borghi, T.; Gusmeroli, R.; Zambra, G.; Oliyink, A.; Fadiga, L.; Spinelli, A.S.; Baranauskas, G. A low-power integrated circuit for analog spike detection and sorting in neural prosthesis systems. In Proceedings of the IEEE Biomedical Circuits and Systems Conference (BioCAS 2008), Baltimore, MD, USA, 20–22 November 2008; pp. 257–260.
28. Andreani, P.; Fard, A. More on the $1/f^2$ Phase Noise Performance of CMOS Differential-Pair LC-Tank Oscillators. *IEEE J. Solid State Circuits* **2006**, *41*, 2703–2712.
29. Bonfanti, A.; Pepe, F.; Samori, C.; Lacaita, A.L. Flicker noise up-conversion due to harmonic distortion in Van der Pol oscillators. *IEEE Trans. Circuit Syst. I* **2012**, *59*, 1418–1430.
30. The quality factor of the tank is approximately determined by the inductor loss at 400 MHz frequency, giving $Q \cong 70$.
31. Samori, C.; Lacaita, A.L.; Villa, F.; Zappa, F. Spectrum folding and phase noise in LC tuned oscillators.*IEEE Trans. Circuit Syst. I* **1998**, *45*, 781–790.
32. Carson, J.R. Notes on the theory of modulation. *Proc. Inst. Radio Eng.* **1922**, *10*, 57–64.
33. Tan, C.; Tjhung, T.; Singh, H. Performance of Narrow-Band Manchester Coded FSK with Discriminator Detection. *IEEE Trans. Commun.* **1983**, *31*, 659–667.
34. Lacaita, A.; Levantino, S.; Samori, C. *Integrated Frequency Synthesizers for Wireless Systems*; Cambridge University Press: Cambridge, UK, 2007.
35. Abdhelhalim, K.; Genov, R. 915-MHz Wireless 64-Channel Neural Recording SoC with Programmable Mixed-Signal FIR Filters. In Proceedings of the 2011 European Solid-State Device Research Conference (ESSCIRC), Helsinki, Finland, 12–16 September 2011; pp. 223–226.

36. Liew, W.-S.; Zou, X.; Lian, Y. A 0.5-V 1.13-μW/channel neural recording interface with digital multiplexing scheme. In Proceedings of the 2011 European Solid-State Device Research Conference (ESSCIRC), Helsinki, Finland, 12–16 September 2011; pp. 219–222.

37. Martinez, J.; Pedreira, C.; Ison, M.J.; Quiroga, R.Q. Realistic simulation of extracellular recordings. *J. Neurosci. Meth.* **2009**, *184*, 285–293.

Chapter 9

Low-Power Photoplethysmogram Acquisition Integrated Circuit with Robust Light Interference Compensation

Jongpal Kim [1], Jihoon Kim [2] and Hyoungho Ko [2,*]

[1]Samsung Electronics Inc., Suwon 16678, Korea

[2]Chungnam National University, Daejeon 34134, Korea

ABSTRACT

To overcome light interference, including a large DC offset and ambient light variation, a robust photoplethysmogram (PPG) readout chip is fabricated using a 0.13-μm complementary metal–oxide–semiconductor (CMOS) process. Against the large DC offset, a saturation detection and current feedback circuit is proposed to compensate for an offset current of up to 30 μA. For robustness against optical path variation, an automatic emitted light compensation method is adopted. To prevent ambient light interference, an alternating sampling and charge redistribution technique is also proposed. In the proposed technique, no additional power is consumed, and only three differential switches and one capacitor are required. The PPG readout channel consumes 26.4 μW and has an input referred current noise of 260 pArms.

Keywords: photoplethysmogram; transimpedance amplifier; ambient light cancellation; automatic offset compensation; automatic emitted light compensation

1. INTRODUCTION

The photoplethysmogram (PPG) is widely used for monitoring cardiovascular parameters such as heartbeat rate, blood vessel stiffness, and pulse transit time blood pressure [1,2,3,4]. In particular, wearable pulse rate sensors based on the PPG have become increasingly popular because of their convenience compared to electrocardiogram sensors. The operation principle of PPG sensors is optical detection of blood volume changes. The sensor system consists of photodiodes (PDs), red, infrared, or green light-emitting diodes (LEDs), and a transimpedance amplifier (TIA). The PPG sensor monitors changes in the light intensity via reflection from the skin or transmission through the tissue [5].

In a PPG monitoring system, light waves emitted from a known and controllable optical source illuminate the fingertip, wrist, or earlobe, and the resultant transmitted or reflected light waves are detected by a PD. A PPG probe is generally worn on the fingertips because a higher signal amplitude can be achieved there than at other sites. Recently, different measurement sites for PPG sensors have been explored extensively for wearable applications, including the ring finger [6], wrist [7], brachia [8], and earlobe [9].

According to the measurement principle, in addition to the signal component of interest, two main interference components are induced by the ambient light intensity and optical path variation. These interferences can be either a direct current (DC) or an alternating current (AC) component of a detected signal. The DC offset of the PPG waveform corresponds to the detected transmitted or reflected optical signal from the tissue and depends on the structure of the tissue and the average blood volume of both arterial and venous blood. The AC component shows changes in the blood volume that occur between the systolic and diastolic phases of the cardiac cycle [5]. A DC offset that depends on the static ambient light intensity and skin color is present in a detected signal. The DC offset is typically a hundred times larger than the pulsatile amplitude [3]. Signal saturation from the large DC offset can be prevented by using a wide dynamic range design, but this results in inefficient power consumption. Therefore, the DC offset should be removed before the detected signal is amplified. Owing to variation in the ambient light intensity or optical path condition between the light source and the detector, the unwanted AC component is present in the detected signal. An ambient light, such as the sun or a fluorescent lamp, can be a noise source because it is unknown and unpredictable.

In previous works, this ambient light was cancelled by a feedback technique with a digital-to-analog converter (DAC) and correction circuit blocks [3] or by an additional analog-to-digital converter (ADC) and a digital signal processor [10]. Other studies also reported a few methods of reducing interference, such as rejecting the DC offset by a low-frequency feedback loop [11], extending the dynamic range using a logarithmic amplification [12], using digital feedback [13], and using an adaptive digital filter [14,15].

This study presents a low-power PPG readout integrated circuit (IC) that is robust to the DC light offset and time-varying interference [16]. Three compensation methods are adopted in this study: (1) a saturation detection and current compensation method to reduce the large DC current offset, (2) ambient light cancellation (ALC) through alternating sampling and charge redistribution, and (3) automatic emitted light compensation (AEC) against optical path variation.

2. CIRCUIT DESIGN

2.1. Top Architecture and Overall Work Flow

A top view of the designed architecture is shown in Figure 1. The LED controller controls the amount of light emitted from the LED. The LED is turned on and off for lower power consumption according to the control signal f_LED. The amount of current passing through the LED is controlled by the LED controller depending on the output signal from a low-pass filter (LPF2). The current signal detected by the PD is converted into a voltage signal in the TIA block, and the ambient light effect is cancelled in the ALC block. A considerable amount of the DC offset from the PD is compensated by the current compensator so that it does not saturate a signal at the ADC input stage. The first low-pass filter (LPF1) has a cutoff frequency of 70 Hz to suppress the ADC noise, and the second low-pass filter (LPF2) has a cutoff frequency of 0.1 Hz to extract the DC component.

2.2. Fully Differential Transimpedance Amplifier (TIA)

A schematic of the fully differential TIA is shown in Figure 2. The PD is operated in photovoltaic mode in a fully differential manner. In photovoltaic mode, a reduced dark current, higher linearity, and lower noise can be obtained than in the photoconductive mode. The fully differential architecture also effectively reduces common mode noises.

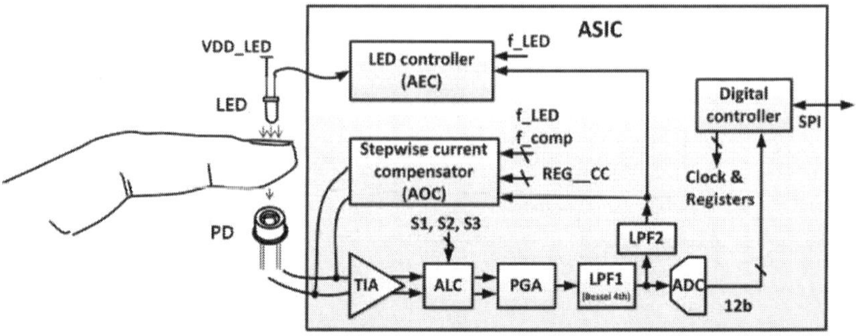

Figure 1. Top block diagram of LED driving and PD detection circuits for PPG application. (TIA: TransImpedance Amplifier; ALC: Ambient Light Cancellation; PGA: Programmable Gain Amplifier; LPF: Low-Pass Filter; PD: PhotoDiode; LED: Light-Emitting Diode; AOC: Automatic Offset Compensation; AEC: Automatic Emitted Light Compensation.)

To reduce the noise level, combined chopper stabilization and power gating is applied. Chopper stabilization is an effective technique for reducing the low frequency flicker noise components [17]. Flicker noise is observed when the state of the transistor is periodically switched from inversion to accumulation by switching the gate bias [18]. We combine the chopper stabilization and switched bias techniques to reduce both the noise and the power consumption. The gate bias voltages, such as CASP, CASN, and BIASN, are periodically switched, and the bias current of the transistors is switched between operating points and off-states; thus, the average current consumption can be reduced while low noise performance is maintained.

The photocurrent generated from the PD, I_{PD}, is modulated by the input chopper (CH1). The modulated photocurrent and two bias currents, I_B, form two differential currents, $I_B + I_{PD}$ and $I_B - I_{PD}$. The differential currents are amplified using a current mirror with a W/L ratio of k. The amplified currents are converted to voltage signals using R_1 and are demodulated to the baseband. The output voltage V_{OUT} can be expressed as in Equation (1).

$$V_{OUT} = 2 \times k \times R_1 \times I_{PD} \quad (1)$$

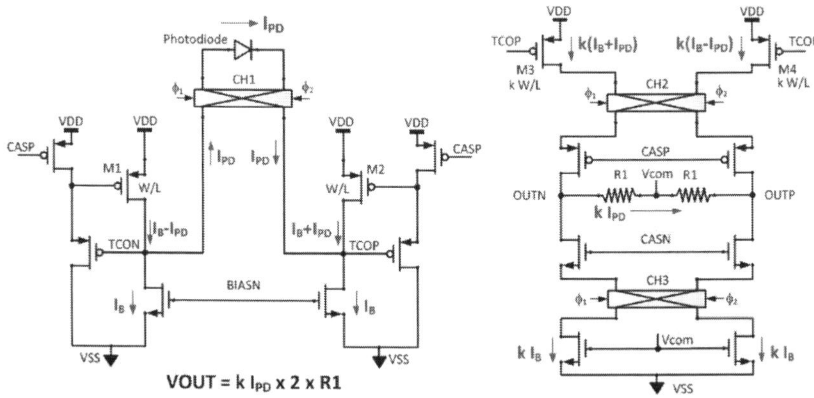

Figure 2. Schematic diagram of fully differential TIA.

2.3. Ambient Light Cancellation (ALC)

The ambient light component of the detected PPG signal can be removed in the ALC block, which is shown in Figure 3. When the LED light turns on, the measured signal corresponding to both the LED light and the ambient light is sampled at the charge reservoir C_r, and when the LED light turns off, the measured signal corresponding to only the ambient light is sampled at another charge reservoir, C_a. Next, charge redistribution occurs to cancel the ambient light component by reconnecting the two charge reservoirs with reversed polarity.

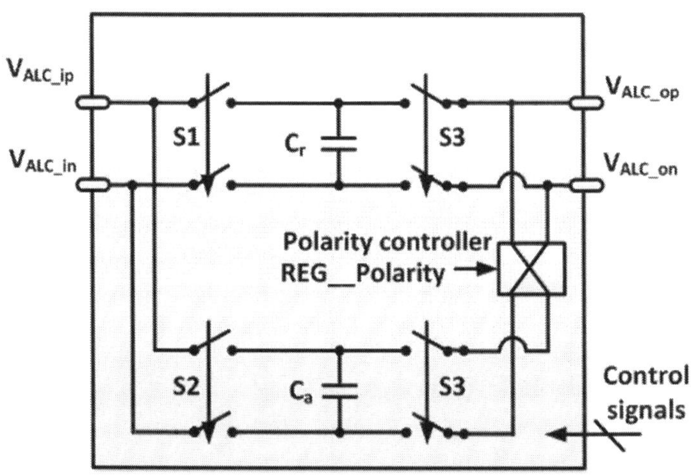

Figure 3. Detailed architecture of the ALC block.

A detailed diagram of the switch operation clock is shown in Figure 4. In Phase 1, Switch S1 is connected, and Switches S2 and S3 are disconnected. During Phase 1, the differential signal $V_{ALC\_ipn\_ph1}$ is sampled in the charge reservoir C_r, and the saved charge can be expressed as follows.

$$Q_r = C_r \times V_{ALC\_ipn\_ph1} \tag{2}$$

In Phase 2, Switch S2 is connected, and Switches S1 and S3 are disconnected. During Phase 2, the differential signal $V_{ALC\_ipn\_ph2}$ is sampled in the charge reservoir C_a, and the saved charge is expressed as follows:

$$Q_a = C_a \times V_{ALC\_ipn\_ph2} \tag{3}$$

In Phase 3, Switch S3 is connected, and Switches S1 and S2 are disconnected. When the control signal REG__Polarity is in the logic state "high", the positive and negative nodes of the reservoir C_r are connected to the negative and positive nodes of the reservoir C_a, respectively. In contrast, when the control signal REG__Polarity is in the logic state "low", the positive and negative nodes of C_r are connected to the positive and negative nodes of C_a, respectively. The polarity controller is used to evaluate the efficacy of the proposed ALC technique.

In this manner, the sampled charges in C_r and C_a are redistributed, and the differential output voltage V_{ALC_opn}at the ALC block is expressed as follows:

$$V_{ALC\_opn} = (C_r \times V_{ALC\_ipn\_ph1} - C_a \times V_{ALC\_ipn\_ph2})/(C_r + C_a) \tag{4}$$

When this technique is implemented, sampling charge reservoirs of the same size will be used; therefore, Equation (4) can be simplified to the following:

$$V_{ALC\_opn} = (V_{ALC\_ipn\_ph1} - V_{ALC\_ipn\_ph2})/2 \tag{5}$$

The sampled signal $V_{ALC\_ipn\_ph1}$ in Phase 1 includes the desired signal $V_{ALC\_ipn\_LED}$ corresponding to the LED light and the undesired signal $V_{ALC\_ipn\_AL}$ corresponding to the ambient light. The signal component from the ambient light should be removed because it is an uncontrollable noise. The sampled signal $V_{ALC\_ipn\_ph2}$ in Phase 2 includes only the undesired signal $V_{ALC\_ipn\_AL}$ because the LED light is turned off. Finally, at the output node of the ALC block, the differential signal without the ambient light interference is obtained:

$$V_{ALC\_opn} = V_{ALC\_ipn\_LED}/2 \tag{6}$$

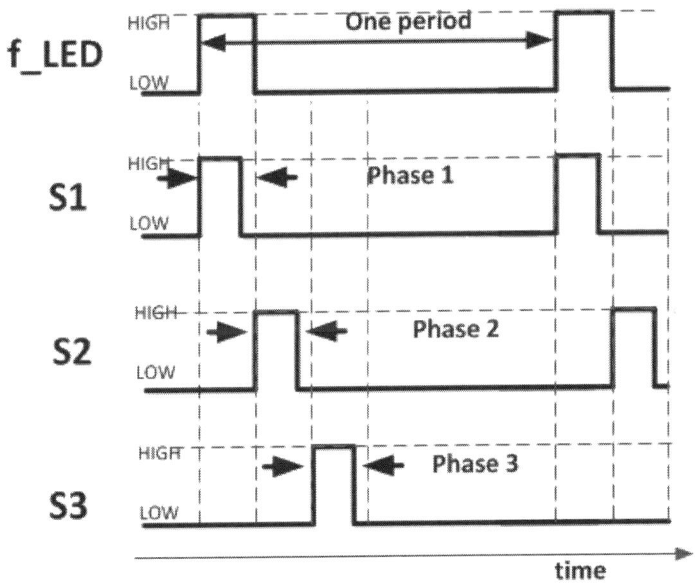

Figure 4. Timing diagram of operation clock for LED power-gating and alternating sampling and charge redistribution.

2.4. Automatic Offset Compensation (AOC)

The large DC offset current compared with the AC pulse current from the PD is compensated using the current compensator (Figure 5). The range comparator generates information about whether the input signal is oversaturated, undersaturated, or not saturated based on comparison with the upper and lower threshold values, which can be adjusted using the registers REG__th_H and REG__th_L. Depending upon this information, the compensator logic increases or decreases the compensation current. Current compensation is implemented using an updated value at the frequency f_comp and is effective when the LED turns on. The current compensator includes an 8-bit current DAC that compensates a DC offset of up to 30 μA.

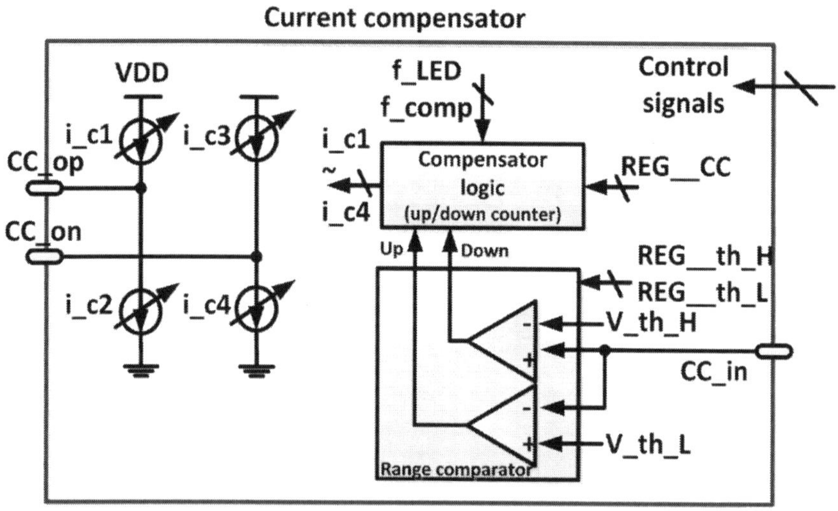

Figure 5. Block diagram of the current compensator block.

2.5. Automatic Emitted Light Compensation (AEC)

Most of the DC current is compensated by the current compensator, and the remaining DC current is used for AEC. Figure 6 shows a block diagram of the AEC loop. The detected and amplified signal that passes through the PD and TIA is filtered by an inverting switched-capacitor low-pass filter (ISC-LPF) with a cutoff frequency of 0.1 Hz to extract the DC component. To reduce the average power consumption of the LED, the LED is turned on and off periodically by turning the drain metal–oxide–semiconductor (MOS) on and off. When the LED turns on, the drain MOS is controlled by the filtered signal from the ISC-LPF. For example, when the detected current signal from the PD decreases, the voltage output of the ISC-LPF increases, thus increasing the LED driving current. The increased LED light leads to an increase in the PD current. Therefore, the input signal level of the ADC is maintained.

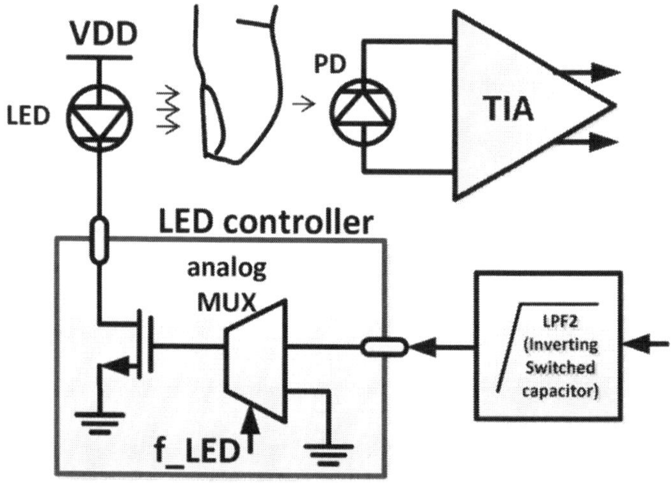

Figure 6. Block diagram of the automatic emitted light compensation (AEC) block.

3. MEASUREMENT RESULTS AND DISCUSSION

The designed circuit is implemented using the 0.13-μm complementary metal–oxide–semiconductor (CMOS) process (Figure 7). The overall size, including two readout channels, the LED driver, the bias circuit, and a clock generator is 6.4 mm × 3.1 mm. Figure 8 shows a photograph of the printed circuit board (PCB) used for performance evaluation. The fabricated PPG readout IC and PPG finger probe are mounted on the PCB. Figure 9a shows the frequency response of the readout channel. The readout channel has a transimpedance gain of 160 kOhm to 3.5 MOhm and a low-pass filter with a cutoff frequency of 70 to 500 Hz, which can be set by the register.Figure 9b shows the noise characteristics of the readout channel, which has a noise level of 260 pArms with a bandwidth of 100 Hz. The readout channel, including the TIA, ALC, PGA, and low-pass filters (LPF1 and LPF2) consumes 22 μA at an operation voltage of 1.2 V.

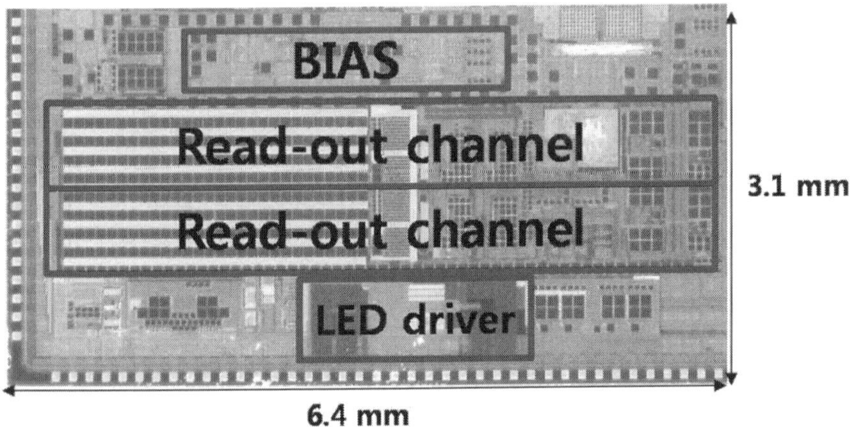

Figure 7. Photograph of chip fabricated by 0.13-µm CMOS process.

Figure 8. Printed circuit board (PCB) used for evaluation.

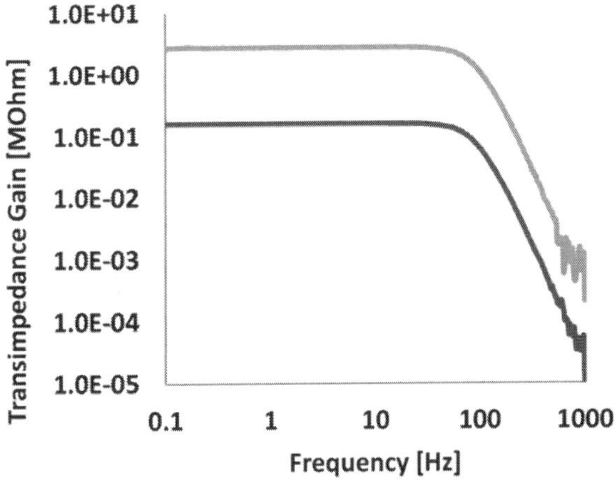

(a)

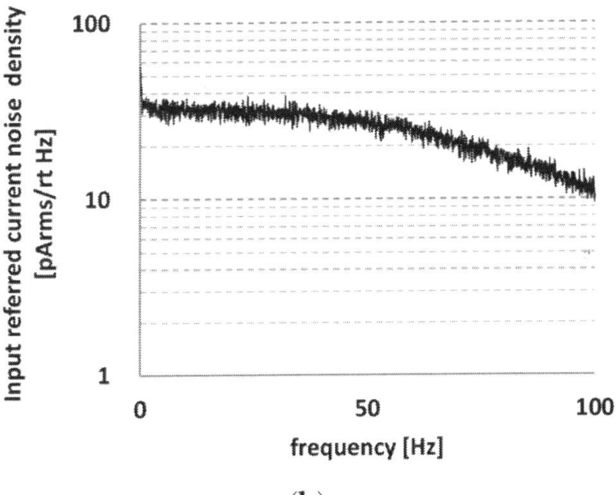

(b)

Figure 9. Transimpedance gain and input referred noise. (**a**) Transimpedance gain with minimum and maximum settings; (**b**) Input referred noise spectrum of readout channel.

To validate the efficacy of the proposed ALC technique, a PPG (Probe 1) is placed on the index finger, and an another PPG (Probe 2) is placed on the middle finger, as shown in Figure 10. PPG Probes 1 and 2 are connected to readout Channels 1 and 2, respectively. Readout Channel 1 is set to work with the ALC and readout Channel 2 is set not to work with the ALC by changing the connection of a polarity controller, as described in Section 2.2. Owing to the variation of light from a fluorescent lamp, the signal from Channel 2 has a peak component that is influenced by the incident ambient light variation, whereas the signal from Channel 1 is not influenced by the ambient light variation. Figure 11 shows that the proposed alternating sampling and charge redistribution technique can eliminate the ambient light interference.

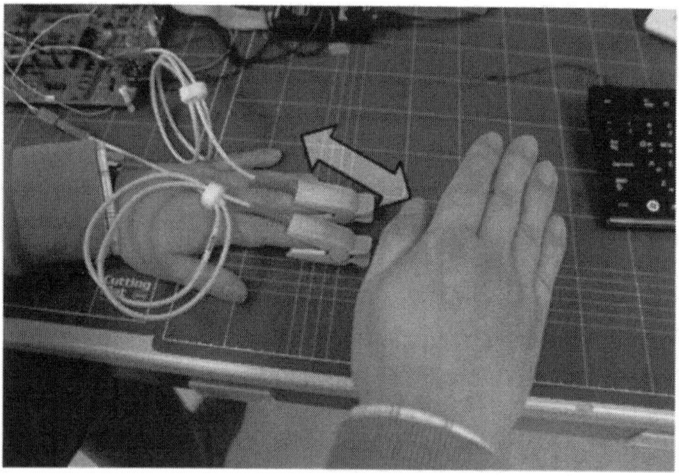

Figure 10. Validation of the ALC method.

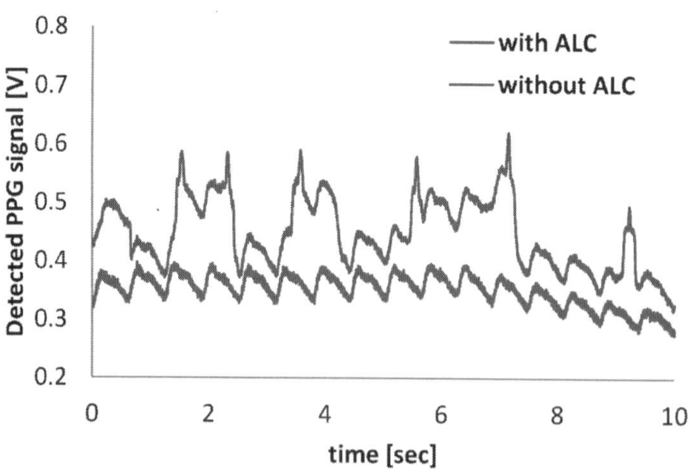

Figure 11. Comparison of output signal with and without the ALC function.

To evaluate the ability of the proposed technique to compensate for the large DC offset current, a combined current consisting of a sine wave of 1 Hz and a DC is applied as an input. As shown in Figure 12, when the monitored signal is saturated, the AOC restores the output signal to the desired range. If the monitored signal is higher or lower than the upper or lower threshold voltage of the range comparator in the AOC block, the up/down counting logic and binary weighted current compensator adjust the output offset to the desired value. DC offset compensation of up to 30 μA is achieved experimentally. The stepwise current compensator based on up/down counter logic is operated with a programmable clock frequency (f_comp) of 0.1 Hz to 2 kHz. A subhertz counter operation clock was used intentionally to obtain the results in Figure 12 to show the transient progress of the AOC function. The transient responses of the AOC can be adaptively controlled by programming the clock frequency of the current compensator.

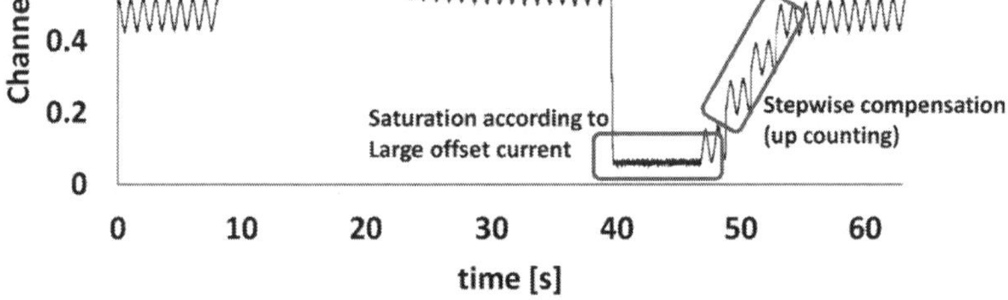

Figure 12. Restoration of saturated signal by the AOC.

Figure 13 shows the response of the AEC function to external interference. Ambient light is used as the external interference, and the ALC function is turned off. As shown in Figure 13, the ambient light is applied initially. When the ambient light is blocked, the output signal goes down and up for a moment. When the ambient light is applied again, the signal returns gradually to the original level.

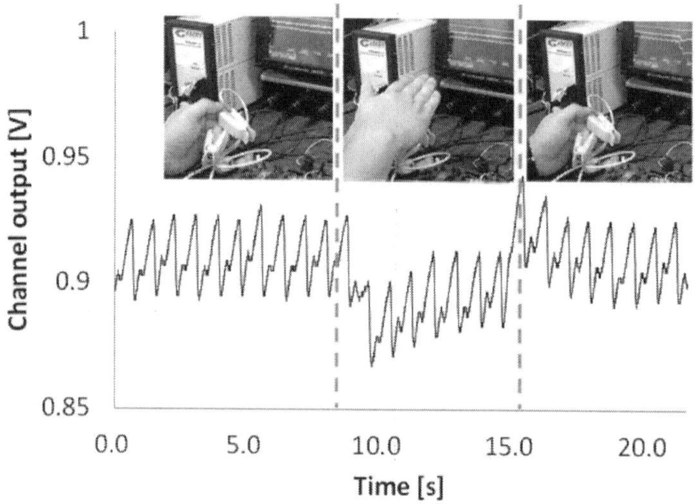

Figure 13. Response of the AEC to artificial interference.

Figure 14 shows the measurement results during slow stepping in place with or without the compensation function. Because the fabricated ASIC is embedded on a large evaluation board wired to instruments, the walking experiment was performed using in-place stepping instead of free walking. In the simultaneously acquired signals shown in Figure 14, the output signal without compensation fluctuated severely; however, a relatively stable output signal with compensation appears under motion artifact conditions.

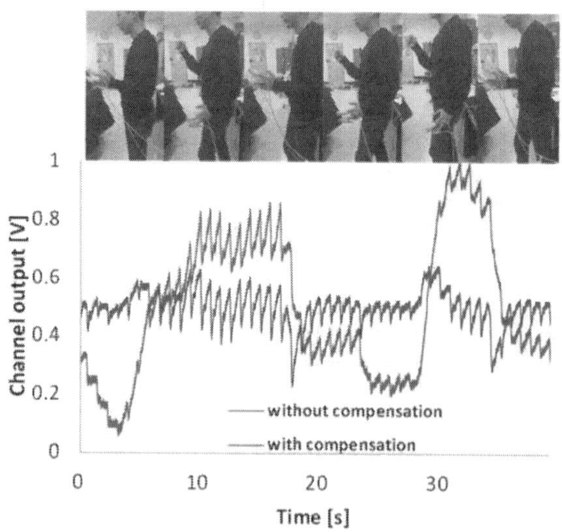

Figure 14. Response of compensation circuit with motion artifact.

Table 1. Comparison to previous works, mainly in power consumption and compensation against interference.

	This Work	[11]	[12]	[13]
Technology [μm]	0.13	0.35	1.5	0.18
Readout channel power consumption (w/o LED power) [μW]	26.4	N/A	160	117
Input referred noise [pArms]	260	N/A	N/A	N/A
Automatic offset compensation (AOC)	Yes	Yes	Yes	Yes
Ambient light cancellation (ALC)	Yes	No	No	No
Automatic emitted light compensation (AEC)	Yes	Yes	Yes	No

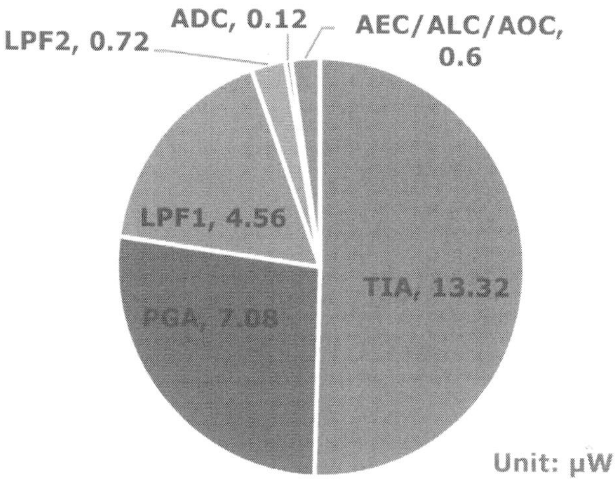

Figure 15. Power breakdown of the presented PPG readout IC.

4. CONCLUSIONS

For a robust PPG application, three compensation methods—An AOC circuit for large DC current offset, ALC through alternating sampling and charge redistribution, and AEC against optical path variation—were implemented, and their efficacy was proved experimentally. To obtain both low-noise and low-power performance, a fully differential TIA with a combined chopper stabilization and switched bias technique was designed. In the ALC circuit, the proposed alternating sampling and charge redistribution technique required only three differential switches and a capacitor. To the best of our knowledge, our proposed ALC technique is the simplest method used to cancel ambient light interference. The transient response of the system can be controlled digitally. Robust PPG acquisition performance with minimized transient response degradation under light interference can be achieved using adaptive digital control from the host microcontroller unit. Performance comparisons to previous works are summarized in Table 1. The power breakdown of the IC is shown in Figure 15. The IC has a low power consumption of 26.4 μW. The additional power consumption for the compensation circuits, *i.e.*, the AEC, ALC, and AOC, is only 0.6 μW. The presented PPG readout IC shows low-power, low-noise, and robust acquisition performance and can be applied to wearable platforms.

ACKNOWLEDGMENTS

This work was partially supported by Chungnam National University.

AUTHOR CONTRIBUTIONS

Jongpal Kim is the first author, and he wrote the draft of the manuscript. He also implemented the ADC and designed the top architecture of the IC. Jihoon Kim designed the digital block of the IC. Hyoungho Ko is the corresponding author, and he designed the analog blocks of the IC.

REFERENCES

1. Bortolotto, L.A.; Blacher, J.; Kondo, T.; Takazawa, K.; Safar, M.E. Assessment of vascular aging and atherosclerosis in hypertensive subjects: Second derivative of photoplethysmogram *versus* pulse wave velocity. *Am. J. Hypertens.* **2000**, *13*, 165–171.
2. Poon, C.; Zhang, Y. Cuff-less and noninvasive measurements of arterial blood pressure by pulse transit time. In Proceedings of the 27th Annual International Conference of the Engineering in Medicine and Biology Society (IEEE-EMBS 2005), Shanghai, China, 1–4 September 2005; pp. 5878–5880.
3. Sant, L.; Fant, A.; Torta, P.; Dorrer, L. A system containing an ambient light and a proximity sensor with intrinsic ambient light rejection. In Proceedings of the 38th European Solid-State Circuit Conference (ESSCIRC 2012), Bordeaux, France, 17–21 September 2012; pp. 97–100.
4. He, D.; Morgan, S.P.; Trachanis, D.; van Hese, J.; Drogoudis, D.; Fummi, F.; Hayes-Gill, B.R. A single-chip CMOS pulse oximeter with on-chip lock-in detection. *Sensors* **2015**, *15*, 17076–17088.

5. Tamura, T.; Maeda, Y.; Sekine, M.; Yoshida, M. Wearable photoplethysmographic sensors—Past and present. *Electronics* **2014**, *3*, 282–302.
6. Rhee, S.; Yang, B.; Asada, H. Artifact-resistant, power-efficient design of finger-ring plethysmographic sensors. *IEEE Trans. Biomed. Eng.* **2001**, *48*, 795–805.
7. Lee, Y.; Shin, H.; Jo, J.; Lee, Y. Development of a wristwatch-type PPG array sensor module. In Proceedings of the IEEE International Conference on Consumer Electronics, Berlin, Germany, 6–8 September 2011; pp. 168–171.
8. Maeda, Y.; Sekine, M.; Tamura, T. Relationship between measurement site and motion artifacts in wearable reflected photoplethysmography. *J. Med. Syst.* **2011**, *35*, 969–976.
9. Poh, M.-Z.; Swenson, N.C.; Picard, R.W. Motion-tolerant magnetic earring sensor and wireless earpiece for wearable photoplethysmography. *IEEE Trans. Inform. Tech. Biomed.* **2010**, *14*, 786–794.
10. AFE4400 Integrated Analog Front-End for Heart Rate Monitors and Low-Cost Pulse Oximeters. Available online: http://www.ti.com/lit/ds/symlink/afe4400.pdf (accessed on 1 November 2015).
11. Wong, A.; Pun, K.; Zhang, Y.; Leung, K. A low power CMOS front-end for photoplethysmographic signal acquisition with robust DC photocurrent rejection. In Proceedings of the 4th IEEE/EMBS International Summer School and Symposium on Medical Devices and Biosensors (ISSS-MDBS 2007), Cambridge, UK, 19–22 August 2007; pp. 53–56.
12. Tavakoli, M.; Turicchia, L.; Sarpeshkar, R. An ultra-low-power pulse oximeter implemented with an energy-efficient transimpedance amplifier. *IEEE Trans. Biomed. Circuits Syst.* **2010**, *4*, 27–38.
13. Winokur, E.; O'Dwyer, T.; Sodini, C.G. A low-power, dual-wavelength photoplethysmogram (PPG) SoC with static and time-varying interferer removal. *IEEE Trans. Biomed. Circuits Syst.* **2015**, *9*, 581–589.
14. Ram, M.; Madhav, K.; Krishna, E.; Komalla, N.; Reddy, K. A novel approach for motion artifact reduction in PPG signals based on AS-LMS adaptive filter. *IEEE Trans. Instr. Meas.* **2012**, *61*, 1445–1457.
15. Peng, F.; Zhang, Z.; Gou, X.; Liu, H.; Wang, W. Motion artifact removal from photoplethysmographic signals by combining temporally

constrained independent component analysis and adaptive filter. *Biomed. Eng. Online* **2014**, *13*.

16. Kim, J.; Lee, T.; Kim, J.; Ko, H. Ambient light cancellation in photoplethysmogram application using alternating sampling and charge redistribution technique. In Proceedings of the 37th Engineering in Medicine and Biology Society (IEEE-EMBC 2015), Milan, Italy, 25–29 August 2015; pp. 6441–6444.
17. Xu, J.; Fan, Q.; Huijsing, J.H.; Van Hoof, C.; Yazicioglu, R.F.; Makinwa, K. Measurement and analysis of current noise in chopper amplifiers. *IEEE J. Solid State Circ.* **2013**, *48*, 581–589.
18. Klumperink, E.A.; Gierkink, S.L.; Van Der Wel, A.P.; Nauta, B. Reducing MOSFET 1/f noise and power consumption by switched biasing. *IEEE J. Solid State Circ.* **2000**, *35*, 994–1001.

Chapter 10

On Design of Memristive Amplifier Circuits

Timur Ibrayev[1], Irina Fedorova[1], Akshay Kumar Maan[2,3], Alex Pappachen James[1*]

[1]School of Engineering, Nazarbayev University, Astana, Republic of Kazakhstan
[2]Griffith University, Brisbane, Australia
[3]Enview R & D Labs, Trivandrum, India

ABSTRACT

Amplifiers are essential building blocks of a majority of the mixed signal circuits that are used in the development of cognitive computing architectures. Their implementation and use is chal- lenged by the second order effects that dominate the MOSFET operations with reduction in the technology size and scale. The ability to program the amplifiers once fabricated becomes an even more challenging problem as it warrants the use of multiple circuit components that lowers cir- cuit performance and in turn outweighs the advantages of generalisation abilities. In this paper, a reconfigurable set of amplifier circuits are proposed based on quantised conductance devices in combination with MOSFET devices. The presented circuits form the basic configurations for the memristor based amplifiers, and show promising performance results in terms of power dissipa- tion, on-chip area and THD values.

Keywords: Resistive Switching, Memristive Device, Quantized Conductance, Common Source Amplifier, Common Drain Amplifier, Differential Amplifier

1. INTRODUCTION

The programmability of analog amplifiers is a challenging, and emerging topic of interest to developing circuits and systems required for futuristic neuromorphic computing architectures [1] [2]. Traditionally, the primary objective of the amplifiers has been to amplify the voltage or current signals and also to match the impedance be- tween circuits [3] - [5]. The ability to amplify signals is critical in sensory signal processing and neuromorphic networks to improve the quality and functionality of the modules. Neuromorphic networks are arranged in layers of neurons that are interconnected in an hierarchical manner, with the ability to transmit and transform the signal from input to output for a meaningful objective [6] - [8]. The ability of the networks to process the signals through learning weights over the connectivity between the network layers requires signal deformations to be minimal [9] - [11]. Increase in the number of the neurons and layers necessitates amplifiers to be integrated into the signal conditioning and processing stages of the neuronal circuit designs [12] - [15]. Since often the signal amplifications required are different for different layers of the network the amplifier designs would need to be tuned to fit into the signaling requirements for each layer [16] [17]. The problem becomes even more challeng- ing when the analog neuron networks are required to grow in size and functionality as the need for number of neurons increases with number of layers and inputs [18] [19], which in turn puts additional burden on the need to ensure further signal conditioning to ensure accuracy and speed of learning.

Reconfigurability of the analog circuits in terms to adjusting the open loop gain and impedances offer the de- signers additional flexibility to improve the reliable operation of multilayered networks [20] - [22]. Similar to neuromorphic networks, another example of application where they are vital is the memory processing and sensing [23] [24]. The analog to digital converters [25] [26] and data read/write circuits [27] make use of amplifiers that require different gains and bandwidths. While the goal is to ensure good performance with minimal change in design, the performance benchmarks drive designers to come up with significantly different circuit topologies for achieving parametric targets of gains, bandwidth and higher noise tolerance under low power conditions [28]. The objective of this paper is to advance the idea of programmability of amplifier circuits with very little modi- fication to circuit topology to ensure the structural and functional reproducibility of the amplifiers across the different layers of networks. To achieve this goal, we use simplistic amplifier configurations that make use of memristor like devices to configure the amplifier parameters. The low area requirements and low leakage cur- rents of memristor like devices followed up with the quantization of conductance with a memristive state enable programmability of the circuit.

2. QUANTISED CONDUCTANCE AMPLIFIER CONFIGURATION

Owing to increased second order effects and power considerations, current chip fabrication techniques are ap- proaching their fundamental limits. Resistive switching devices such as memristors offer a promising alternative as large scale, highly dense, programmable electronic memory network components [29]. Resistive switching enables quantised conductance by reversible formation and dissolution of nanometre-scale conductive filaments, which constrain the motion of electrons. Quantum effects begin to dominate and electron transport is quantised in multiples of the fundamental quantum of conductance $G_o\left(=2e^2/h\right)$ where e is the charge of electron and is Planck's constant) [29] , thus leading to the resistive switching phenomenon where an external electric field coupled with the previous resistance state of the device determine the new resistance state.

The physical model of the memristor from [30] [31] shown in Figure 1 consists of a two-layer thin film size ($D \approx 10nm$) of TiO_2, sandwiched between platinum contacts. One of the layers doped with oxygen vacancies behaves as a semiconductor. The second, undopedregion, has an insulating property. As a consequence of quan- tised conductance process described above, the width w of the doped region is modulated depending on the amount of electric charge passing through the memristor. With electric current passing in a given direction, the boundary between the two regions is moving in the same direction. The total resistance of the memristor, M, is a sum of the resistances of the doped and undoped regions,

$$M\left(x\right)=R_{ON}x+R_{OFF}\left(1-x\right), \quad (1)$$

Where

$$x=\frac{w}{D}\in\left(0,1\right) \quad (2)$$

is the width of the doped region, referenced to the total length D of the TiO_2 layer, and R_{OFF} and R_{ON} are the values of the memristor resistance for $w=0$ and $w=D$. The ratio of the two resistances is usually given as 10^2-10^3 [30].

The quantised conductance type [29] memristive devices offer different levels of conductance values within its state, that enables the discretized control of resistor values. Since resistors play a major role in the biasing and control of MOSFET amplifier parameters, the replacement of resistors with memristors can offer several advan- tages as compared to a resistor based design. The resistors in integrated circuits are implemented as semiconductor

resistors or as MOSFET pseudo-resistors that have a disadvantage of large reverse currents undesired in a resis- tor. In addition, a large resistance value is often associated with large area of semiconductor resistor or limited by the threshold voltage control in case of MOSFET pseudo-resistors. In contrast, quantised conductance me- mristors do not have the issue of reverse leakage currents and do not require to be scaled to achieve higher re- sistor values. The presented reconfigurable topology of common source, common drain and differential amplifier is built by using a combination of quantised conductance memristor like devices and MOSFET devices.

2.1. Common Source Amplifier

Figure 2 shows the common source amplifier that uses memristors M_1 and M_2 to form the memristive po- tential divider circuit to bias the transistor T, and M_D controls the gain of the amplifier, and also plays a role in the output impedance of the amplifier. The open loop voltage gain of the amplifier is given as:

$$A_{vo} = g_m\left(r_o \| M_D\right), \tag{3}$$

where r_o is the internal output resistance and g_m is the transconductance of the MOSFET. Clearly, the ability to program the values of the M_D enables the control of the amplifier open loop gain. The input resistance of the common source amplifier r_{in} is defined in terms of the memristors M_1 and M_2 and is expressed in the Equation (4).

$$r_{in} = M_1 \| M_2 \tag{4}$$

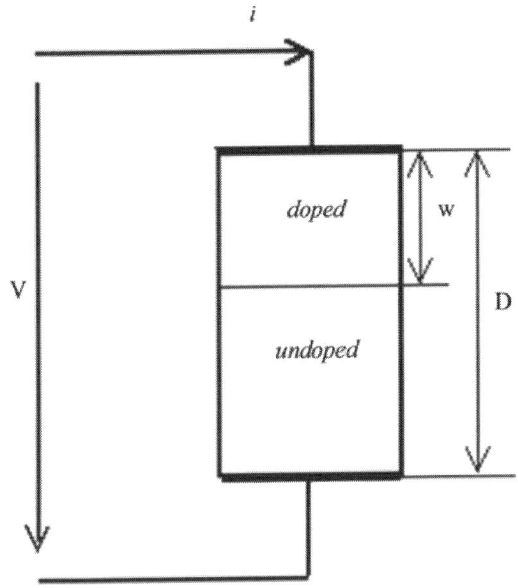

Figure 1. Memristor model [31].

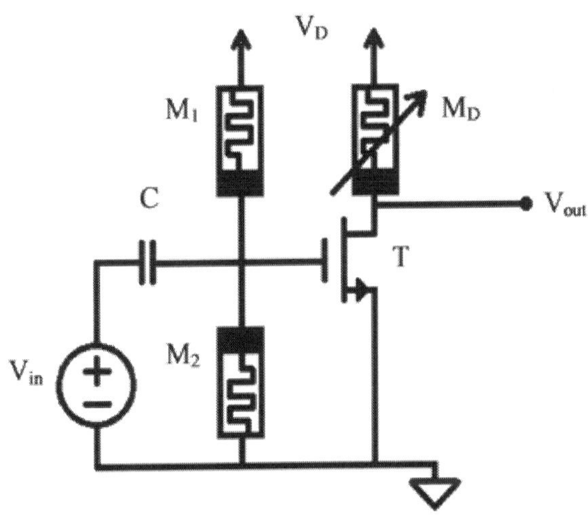

Figure 2. Circuit diagram of common source amplifier.

The output resistance of the circuit is given as:

$$r_{out} = r_o \| M_D \tag{5}$$

2.2. Common Drain Amplifier

Circuit of the common drain amplifier could be seen on Figure 3, where two memristors M_1 and M_2 are used in the same way as it was for the common source amplifier and M_S is the variable memristance that is used for the control of the gain of the amplifier. The gain of the amplifier for the infinite output resistance r_o is defined as follows:

$$A_{vo} = \frac{M_1 \| M_2}{M_s + M_1 \| M_2} \frac{M_S}{r_s + M_S}, \tag{6}$$

where r_s is the reciprocal of the transconductance of the MOSFET g_m.

Equation (6) shows that the gain of the common drain amplifier could be controlled by changing the values of the M_S. Similar to the common source configuration the output resistance of the common drain is given as:

$$r_{in} = M_1 \| M_2 \tag{7}$$

The output resistance of the circuit is defined as:

$$r_{out} = r_s \| M_S \tag{8}$$

2.3. Differential Amplifer

Figure 4 represents the differential amplifier configuration with the variable memristive element M_{SS}. Tran- sistors T_1 and T_2 as well as the other two transistors T_3 and T_4 are in parallel.

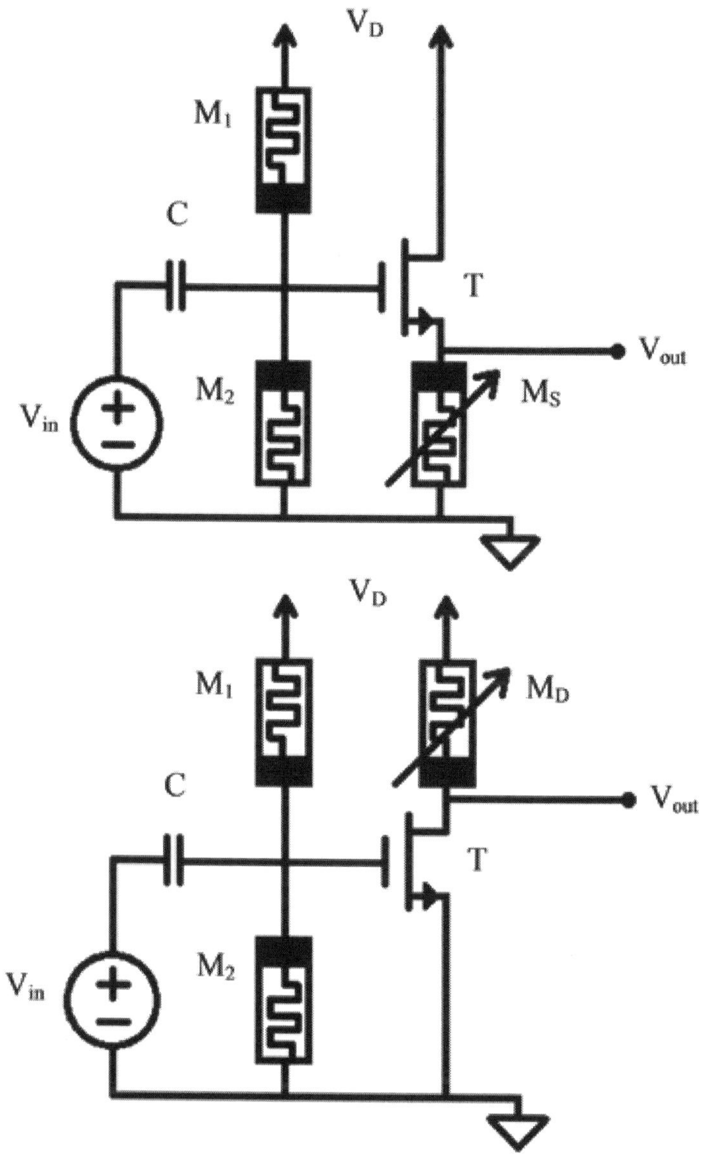

Figure 3. Circuit diagram of common drain amplifier.

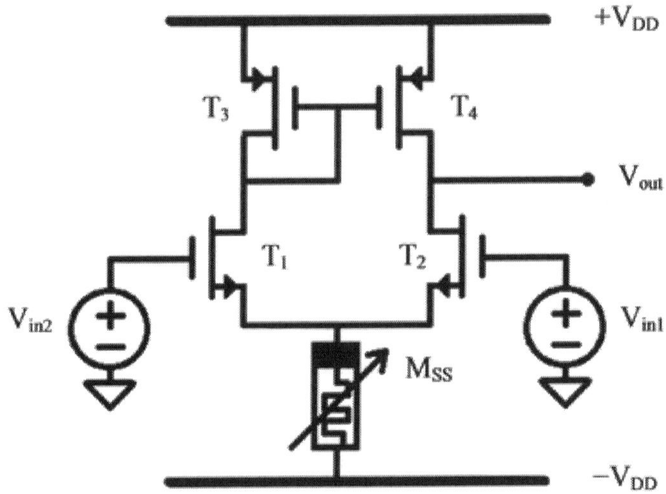

Figure 4. Circuit diagram of differential amplifier.

Common mode gain A_{CM} for the differential pair is expressed as:

$$A_{CM} = \frac{\Delta V_{out}}{\Delta V_{in,CM}}, \tag{9}$$

where ΔV_{out} is the single-ended output parameter and $\Delta V_{in,CM}$ is the common mode input change. It could be further shown that:

$$A_{CM} \approx \frac{-\frac{1}{2g_{t3,4}} \Big\| \frac{m_{o3,4}}{2}}{\frac{1}{2g_{t1,2}} + M_{SS}} = \frac{-1}{1+2g_{t1,2}M_{SS}} \frac{g_{t1,2}}{g_{t3,4}}, \tag{10}$$

where $g_{m3,4}$ and $m_{o3,4}$ are the transconductance and output resistance of the transistors T_3 and T_4 respec- tively and $g_{m1,2}$ is the transconductance of the transistors T_1 and T_2.

3. EXPERIMENTAL RESULTS

The quantised conductance device offers different resistor values in its working states, and we make use of this to adjust the gain of the amplifiers. The ability of the memristor like devices to be programed to different logic state enable the practical realisation of these amplifiers as tunable open loop amplifiers within integrated circuits paradigm. Our quantised conductance memristive device offer resistances of 1.72 kΩ, 1.99 kΩ, 2.15 kΩ, 3.23 kΩ, 6.45 kΩ and 8.60 kΩ. The simulations were done using BSIM models and memristors SPICE model for si- mulating the quantized conductance devices. The parameters from the IBM process technology for different tech- nology size were used for the BSIM models to emulate realistic implementations.

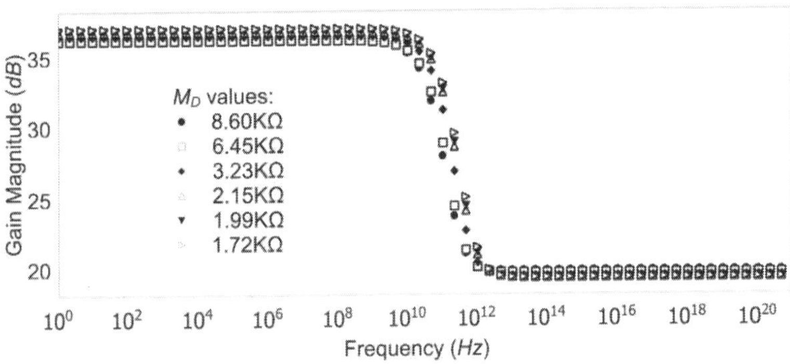

(a)

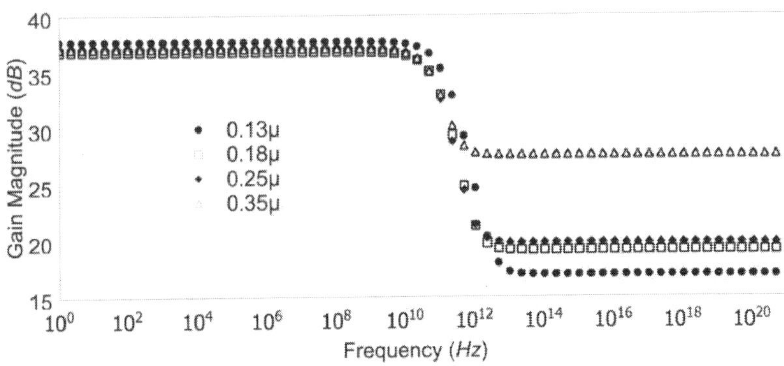

(b)

Figure 5. (a) Graph for common source amplifier showing volatge gain for various values of M_D; (b) Graph for common source amplifer showing voltage gain for various technologies at memristor value of M_D = 1.72 kΩ.

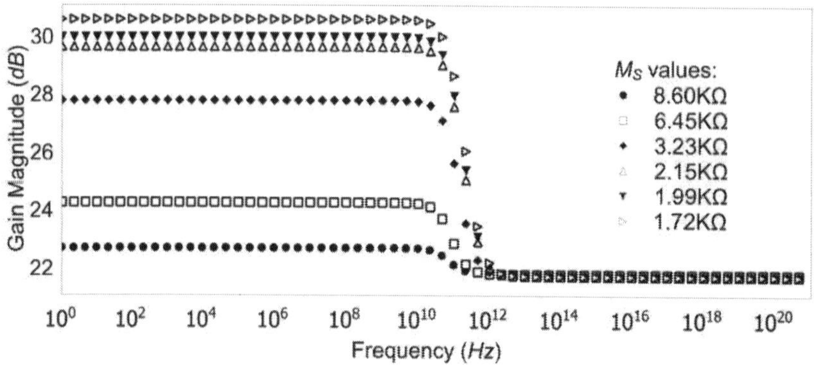

(a)

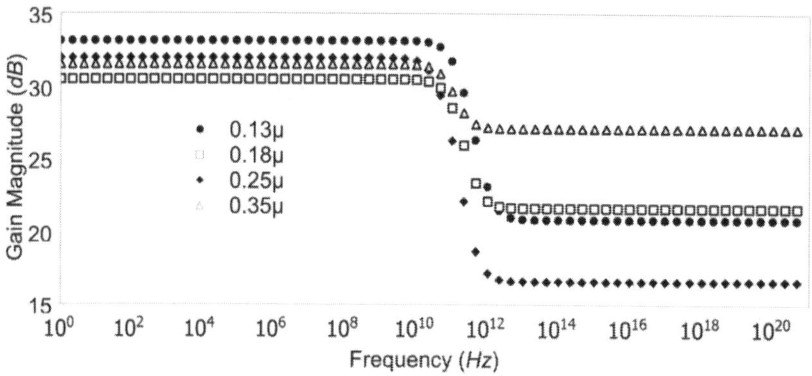

(b)

Figure 6. (a) Graph for common drain amplifier showng volatge gain for various values of M_S; (b) Graph for common drain amplifer showing voltage gain for various technologies at memristor value of M_S = 1.72 kΩ.

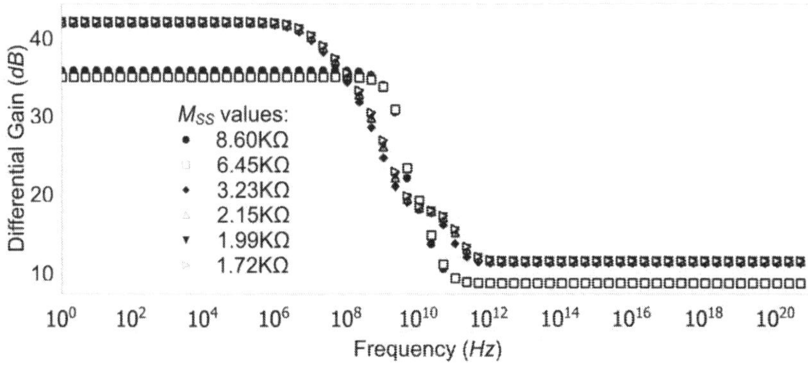

(a)

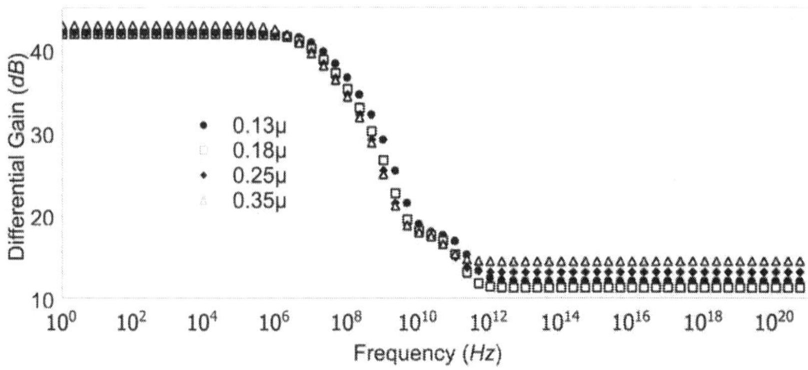

(b)

Figure 7. (a) Graph for differential amplifier showng volatge gain for various values of M_{SS}; (b) Graph for differential amplifer showing voltage gain for various technologies at memristor value of M_{SS} = 1.72 kΩ.

Figure 5 (a) shows the frequency response of the common source amplifier for discrete switching levels of memristive device for 0.18 μ technology of the transistor. The ability to change the resistance of the memristive device allows controlling the gain of the amplifier. Figure 5 (b) shows the impact of technology scaling on the frequency response for an example M_D = 1.72 kΩ. Even with a large change in technology scales the bandwidth and gain is not compromised ensuring practical realisability in wide range of applications.

Figure 6(a) shows the change in gain magnitude of the common drain amplifier for the range of frequencies using different values of memristance for the 0.18 μ technology. Similar to the common source configuration, it is possible to control the voltage gain magnitude of common drain amplifier by implementation of memristor devices. Figure 6(b) shows the voltage gain magnitude at Ms = 1.72 kΩ for the technologies having different scales. It can be seen that it is possible to have similar effect of memristor device despite the variations in tech- nology scales.

Figure 7(a) shows frequency response of the differential gain magnitude for various memristor device values Mss applied with the 0.18 μ technology of the transistors. The variations in the value of memristor device make it possible to alternate the value of the differential gain for the differential amplifier. Figure 7(b) shows the ef- fect of technology scaling on the differential gain magnitude at the Mss = 1.72 kΩ. Because of the similarity in gain magnitude and bandwidth, it is possible to have practical implementation of memristor devices in differen- tial amplifiers for different purposes.

Figure 8 shows the dependency of output resistance on memristance values for three different configurations. It could be observed that the results are in agreement with the corresponding equations describing behaviour of the amplifiers.

The circuit performance in terms of the area required for implementation of the circuit, power consumption and Total Harmonic Distortion (THD) for each of the three configurations are presented in Table 1. Area for common source, common drain and differential amplifiers were calculated taking into account each transistor and memristor of the circuit. These values are lower than that uses semiconductor resistors or pseudo-resistors of MOSFETs. In addition, the simplicity in the design without the need to increase the circuit complexity makes it a useful alternative in mainstream circuit design.

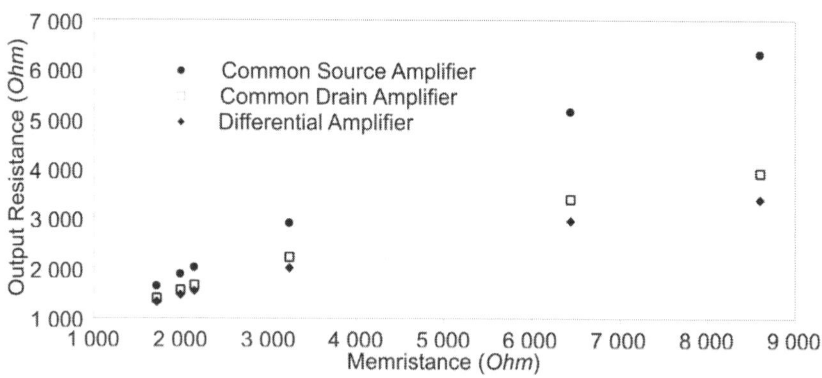

Figure 8. Graph showing output resistance versus memristance values for each of three amplifer.

Table 1. The table indicating the performance measures for different ampli- fier configurations able type.

Configuration	Quantised Conductance Memristors		
	Area (pm^2)[a]	Power (μW)[b]	THD (%)
Common Source	00.11	1292.778	3.106[c]
Common Drain	00.11	1268.492	0.577[c]
Differential	32.90	1805.512	0.601[d]

[a]Assuming memristor dimensions to be 185 nm × 112 nm; [b]Values are taken at M = 1.72 kΩ;[c]Input is sine signal (amplitude = 1 V, frequency = 1 kHz); [d]Input is sine signal (amplitude = 10 mV, frequency = 1 kHz).

4. CONCLUSION

In this paper, we presented basic integrated amplifier circuits that is programmable for output signal gains using memristors like quantised conductance devices and MOSFET. The results of the simulations showed the im- proved performance of the common source, common drain and differential amplifiers in terms of the area, power and THD when implemented using the memristor elements instead of the conventional resistors or pseudo-resis- tors. The idea of controlling the resistance of the amplifier in an integrated circuit and the use of memristor ele- ment in the amplifier design is the new ideas suggested in this work. The proposed approach can have a wide- spread use in the implementation of analog amplifier circuits in sensory signal processing circuits, analog neu- romorphic networks, and memory driven circuits and systems.

REFERENCES

1. Rajendran, J., Manem, H., Karri, R. and Rose, G.S. (2010) Memristor Based Programmable Threshold Logic Array. Proceedings of the 2010 IEEE/ACM International Symposium on Nanoscale Architectures, Anaheim, 17-18 June 2010, 5-10.

2. Ng, K.A. and Chan, P.K. (2005) A CMOS Analog Front-End IC for Portable EEG/ECG Monitoring Applications. IEEE Transactions on Circuits and Systems I: Regular Papers, 52, 2335-2347.
3. Mallinson, M. and Spitalny, P. (1993) Programmable Gain Amplifier. US Patent No. 5233309. US Patent and Trademark Office, Washington DC.
4. Hall, T.S., et al. (2005) Large-Scale Field-Programmable Analog Arrays for Analog Signal Processing. IEEE Transactions on Circuits and Systems I: Regular Papers, 52, 2298-2307.
5. Harrison, R.R., Bragg, J.A., Hasler, P., Minch, B.A. and Deweerth, S.P. (2001) A CMOS Programmable Analog Memory- Cell Array Using Floating-Gate Circuits. IEEE Transactions on Circuits and Systems II: Analog and Digital Signal Processing, 48, 4-11.
6. Türel, Ö. and Likharev, K. (2003) Cross Nets: Possible Neuromorphic Networks Based on Nanoscale Components. In- ternational Journal of Circuit Theory and Applications, 31, 37-53.
7. Afifi, A., Ayatollahi, A. and Raissi, F. (2009) STDP Implementation Using Memristive Nanodevice in CMOS-Nano Neuro- morphic Networks. IEICE Electronics Express, 6, 148-153.
8. Afifi, A., Ayatollahi, A. and Raissi, F. (2009) Implementation of Biologically Plausible Spiking Neural Network Models on the Memristor Crossbar-Based CMOS/Nano Circuits. European Conference on Circuit Theory and Design, ECCTD 2009, Antalya, 23-27 August 2009, 563-566.
9. Turel, O., Lee, J.H., Ma, X. and Likharev, K.K. (2004) Nanoelectronic Neuromorphic Networks (Crossnets): New Re- sults. 2004 IEEE International Joint Conference on Neural Networks, Budapest, 25-29 July 2004.
10. Likharev, K., Mayr, A., Muckra, I. and Türel, Ö. (2003) CrossNets: High-Performance Neuromorphic Architectures for CMOL Circuits. Annals of the New York Academy of Sciences, 1006, 146-163. Türel, Ö., Lee, J.H., Ma, X. and Likharev, K.K. (2004) Neuromorphic Architectures for Nanoelectronic Circuits. International Journal of Circuit Theory and Applications, 32, 277-302.
11. Alspector, J. (1989) Neuromorphic Learning Networks. US Patent No. 4874963.
12. Indiveri, G., Chicca, E. and Douglas, R.J. (2009) Artificial Cognitive Systems: From VLSI Networks of Spiking Neurons to Neuromorphic Cognition. Cognitive Computation, 1, 119-127.

13. Schemmel, J., Fieres, J. and Meier, K. (2008) Wafer-Scale Integration of Analog Neural Networks. IEEE International Joint Conference on Neural Networks, IJCNN 2008, Hong Kong, 1-8 June 2008, 431-438.
14. Zacek, K. and Nikolic, L. Analog Neural Networks. Technical University in Brno, 611.
15. Widrow, B., Rumelhart, D.E. and Lehr, M.A. (1994) Neural Networks: Applications in Industry, Business and Science. Communications of the ACM, 37, 93-105.
16. Boser, B.E., Sackinger, E., Bromley, J., Le Cun, Y. and Jackel, L.D. (1991) An Analog neural Network Processor with Programmable Topology. IEEE Journal of Solid-State Circuits, 26, 2017-2025.
17. Schemmel, J., Hohmann, S., Meier, K. and Schürmann, F. (2004) A Mixed-Mode Analog Neural Network Using Cur- rent-Steering Synapses. Analog Integrated Circuits and Signal Processing, 38, 233-244.
18. Adhikari, S.P., Yang, C., Kim, H. and Chua, L.O. (2012) Memristor Bridge Synapse-Based Neural Network and Its Learn- ing. IEEE Transactions on Neural Networks and Learning Systems, 23, 1426-1435.
19. Xiang, L., Yang, Z., Shujuan, W. and Guofu, Z. (2011) A Method for Analog Circuits Fault Diagnosis by Neural Network and Virtual Instruments. 2011 3rd International Workshop on Intelligent Systems and Applications (ISA), Wuhan, 28-29 May 2011, 1-5.
20. Stoica, A., Keymeulen, D., Zebulum, R., Thakoor, A., Daud, T., Klimeck, Y., et al. (2000) Evolution of Analog Circuits on Field Programmable Transistor Arrays. The 2nd NASA/DoD Workshop on Evolvable Hardware, Palo Alto, 13- 15 July 2000, 99-108.
21. Giannini, V., Craninckx, J., D'Amico, S. and Baschirotto, A. (2007) Flexible Baseband Analog Circuits for Software- Defined Radio Front-Ends. IEEE Journal of Solid-State Circuits, 42, 1501-1512.
22. El Gamal, A. and Eltoukhy, H. (2005) CMOS Image Sensors. IEEE Circuits and Devices Magazine, 21, 6-20.
23. Cabric, D., Mishra, S.M. and Brodersen, R.W. (2004) Implementation Issues in Spectrum Sensing for Cognitive Radios. Conference Record of the 38th Asilomar Conference on Signals, Systems and Computers, 1, 772-776.
24. Le, B., Rondeau, T.W., Reed, J.H. and Bostian, C.W. (2005) Analog-to-Digital Converters. IEEE Signal Processing Magazine, 22, 69-77.
25. Overney, F., Rufenacht, A., Braun, J., Jeanneret, B. and Wright, P.S. (2011) Characterization of Metrological Grade Analog-to-Digital

Converters Using a Programmable Josephson Voltage Standard. IEEE Transactions on Instrumentation and Measurement, 60, 2172-2177.

26. Fukami, S., Ishiwata, N., Numata, H., Ohshima, N., Sugibayashi, T. and Suzuki, T. (2011) Magnetic Memory Cell, Magnetic Random Access Memory, and Data Read/Write Method for Magnetic Random Access Memory. US Patent No. 7929342.
27. McConaghy, T., Palmers, P., Steyaert, M. and Gielen, G.G. (2011) Trustworthy Genetic Programming-Based Synthesis of Analog Circuit Topologies Using Hierarchical Domain-Specific Building Blocks. IEEE Transactions on Evolutionary Computation, 15, 557-570.
28. Mehonic, A., Vrajitoarea, A., Cueff, S., Hudziak, S., Howe, H., Labbé, C., et al. (2013) Quantum Conductance in Silicon Oxide Resistive Memory Devices. Scientific Reports, 3, Article No. 2708.
29. Biolek, Z., Biolek, D. and Biolková, V. (2009) SPICE Model of Memristor with Nonlinear Dopant Drift. Radioengineering, 18, 210-214.
30. Strukov, D.B., Snider, G.S., Stewart, D.R. and Williams, R.S. (2008) The Missing Memristor Found. Nature, 453, 80- 83.

Chapter 11

Design of Network Architecture Using Mobile Gateways for DLNA Devices in Wide Area Networks

Hiroaki Goto, Hideki Shimada, Kenya Sato

Department of Information Systems Design, Doshisha University, Kyoto, Japan

ABSTRACT

The prevalence of information appliances supporting DLNA (Digital Living Network Alliance) such as televisions, recorders, and mobile phones has made it possible to share digital contents (e.g. videos, music and pictures) among appliances connected to a local network. However, DLNA does not let you share contents over different networks via the Internet. In this paper, we propose a network architecture where we adopt our SOAP method to mobile devices and use them as mobile gateways to consume digital contents from remote networks. We also confirm its practicality with a prototype.

Keywords: Network Architecture; DLNA; Mobile Network; Home Appliances

1. INTRODUCTION

Recently, information appliances conforming to the DLNA (Digital Living Network Alliance) Guidelines [1,2] such as televisions, recorders and mobile phones are gaining popularity. The DLNA guidelines define a protocol and a set of features to let information appliances of various vendors communicate with each other without troubling the user with complex configurations. This lets the user, for instance, browse videos, pictures and music stored in a recorder and play them back on her television via a local network [3].

Also, with the prevalence of mobile PCs, mobile phones and smartphones, there is an increasing demand to use information appliances away from home. For instance, one could be at a friend's house and want to watch a movie stored on a recorder at home. One could be outside and want to listen to music stored on a PC at home using a mobile device. Interconnectivity not only among devices on a local network but also on remote networks is expected to gain popular use [4].

However, because the DLNA Guidelines only assume usage contained in local networks, DLNA-conformant appliances residing in different networks cannot communicate with each other (we call this feature Wide Area DLNA Connectivity).

To achieve Wide Area DLNA Connectivity, we have proposed a method (the SOAP method) [5,6] in which a home gateway is equipped with a SOAP (Simple Object Access Protocol [7-9]) Web service. Others have proposed methods employing similar gateway facilities to achieve Wide Area DLNA Connectivity. However, existing methods only enable Wide Area DLNA Connectivity among networks that have home gateways installed.

The aim of this paper is to expand on our SOAP method to achieve Wide Area DLNA Connectivity with networks where home gateways are not (or cannot be) installed. This is done by using mobile devices as gateways to communicate with gateways at home.

2. WIDE AREA DLNA CONNECTIVITY

2.1. Overview

As networked information appliances gain popularity, demands to use home information appliances away from home via the Internet are growing. For instance, one could be at a friend's house and want to play a movie stored on a recorder at home. One could be outside and want to listen to music stored on a PC at home using a mobile device. Interconnectivity not only among devices on a local network but also on remote networks is expected to gain popular use. In this paper, we discuss how to share digital contents among DLNA-conformant appliances residing on different remote networks (we call this feature Wide Area DLNA Connectivity).

2.2. Challenges

Because the DLNA Guidelines only assume usage contained in local networks, it does not support Wide Area DLNA Connectivity natively. The following challenges must be addressed:

1) M-Search, used to discover DLNA-conformant devices, sends a multicast to 239.255.255.250:1900. Because this message cannot reach over the Internet, it cannot discover devices on remote networks.
2) When manipulating devices on the local network, DLNA-conformant devices send UPnP messages based on their local IP addresses. These messages cannot reach devices on remote networks.
3) Many household broadband routers use NAT (Network Address Translation). These routers cannot route messages sent from outside the local network to the desired device on the local network.

3. EXISTING RESEARCHES AND THEIR ISSUES

3.1. Existing Researches

Let us discuss existing researches that aim to resolve the challenges described in Section 2.2 and achieve Wide Area DLNA Connectivity.

3.1.1. SOAP Method

The SOAP method requires that home gateways be placed at the borders of every network that is to participate in DLNA Wide Area Connectivity. These home gateways implement SOAP, the protocol to call remote data and services with HTTP and XML-based messages. The home gateways take the UPnP messages sent and received by DLNA-conformant devices, encapsulate them in SOAP messages, and transfer them over the Internet, thus enabling Wide Area DLNA Connectivity.

3.1.2. Wormhole Devices

Wormhole Devices [10,11] are placed and connected to broadband routers on different networks. These Wormhole Devices encapsulate UPnP messages in SIP messages and transfer them via an SIP server to achieve Wide Area DLNA Connectivity. Moreover, the DMSs and the DMPs involved can communicate as if they were on the same local network, so they do not need to be changed in design or implementation.

3.1.3. VPN Method

The VPN method [12,13] requires that home gateways and home servers be installed in each of the networks. The home gateways are to be placed at the borders of the networks, and the home servers are connected to their local home gateways. The home gateways perform secure communications with each other with IPSec while the home servers generate virtual DLNA-conformant devices that act as delegates of the remote DLNA devices. The gateways and servers provide secure Wide Area DLNA Connectivity.

3.2. Issues with Existing Researches

The researches described above have enabled us to link DLNA-conformant devices on different remote networks. DLNA-conformant devices are expected to gain more popularity as mobile phones, smartphones, STBs (set-top boxes) and printers adopt the DLNA Guidelines. Hence, it is imperative that Wide Area DLNA Connectivity among a variation of networks and devices is achieved. However, the existing researches have limited capabilities due to these issues:

1) They require that home gateways be placed at each and every one of the networks that are to participate in Wide Area DLNA Connectivity. High installation costs are introduced when setting up new networks to participate.

2) Many public settings such as hotels or restaurants make it impossible to set up home gateways. Wide Area DLNA Connectivity cannot be achieved in these settings.

3) They only support DLNA Guidelines version 1.0, which only supports 2-Box Pull. Hence, the devices that support DLNA Guidelines version 1.5 cannot fully utilize their 3-Box capabilities.

4. PROPOSED METHOD

4.1. Overview

In this paper, we expand on our SOAP method to resolve the issues discussed in Section 3.2 and achieve Wide Area DLNA Connectivity. We propose a network architecture that uses a mobile gateway in addition to a home gateway. The mobile gateways run on mobile devices with wireless LAN capabilities, such as notebook PCs, mobile phones and smartphones.

This setup has the following advantages. First, users do not have to set up home gateways at networks away from home, and installation costs are eliminated. Second, it enables Wide Area DLNA Connectivity anywhere wireless LAN is available. Third, because the proposed method uses the mobile gateway as a controller, devices conforming to the DLNA Guidelines version 1.5 can be used in 3-box setups. These advantages resolve all three issues discussed in Section 3.2.

Figure 1 is a schematic diagram of the proposed network architecture. The home gateway is placed either at the border of the home network or behind its broadband router. The mobile gateway uses wireless LAN to connect to whatever network is available.

4.2. Functioning of the Proposed Method

4.2.1. Design Requirements

In designing the functions for our proposed method, we must consider two key characteristics. One is that it uses mobile devices to run gateways on, whereas existing methods use PCs or broadband routers. Therefore, we must take into account the extra power consumption introduced by its processing load, for battery lives of mobile devices are limited. Another is that the proposed method supports DLNA Guidelines version 1.5. Therefore, it must provide 3-box setups.

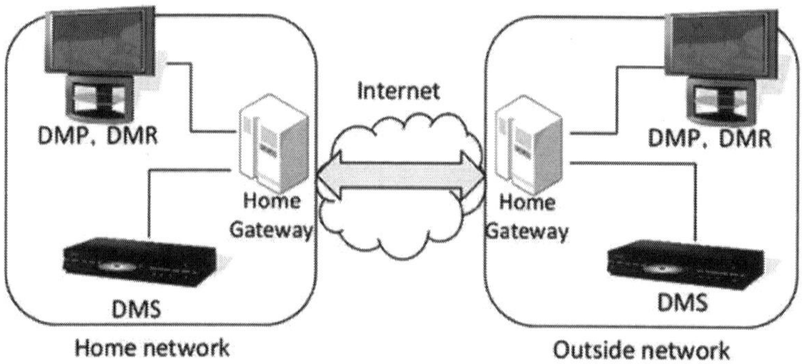

(a) Current network architecture (w/gateway)

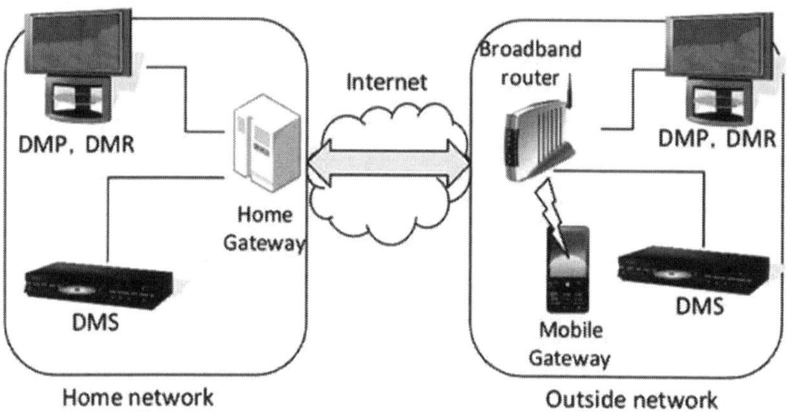

(b) Proposed network architecture (w/gateway)

Figure 1. Network architecture of DLNA connection in wide area.

The proposed method requires the functions described below to achieve Wide Area DLNA Connectivity with mobile gateways.

4.2.2. Functionality of the Home Gateway

Let us describe the functionality of the home gateway.

4.2.2.1. UPnP Message Relay

This functionality attacks issues 1) and 2) described in Section 2.2. First, the mobile gateway encapsulates UPnP messages in SOAP messages and sends them to the home gateway. The home gateway then decapsulates the UPnP message and relays it to the desired DLNA-conformant device on the local network. Messages from the local device to the mobile gateway are relayed similarly. This relay functionality lets devices discover, retrieve information on and manipulate other devices on remote networks.

4.2.2.2. Port Mapping

This functionality attacks issue 3) described in Section 2.2. We use UPnP-IGDs (Internet Gateway Devices, or broadband routers supporting UPnP) to dynamically map ports. This is done by sending AddPortMapping messages to the router from the home gateway. Port mapping lets local DLNA-conformant devices receive messages over the Internet.

4.2.2.3. Proxy Controller

The proxy controller is a local representation of the remote mobile gateway. The proxy controller acts as if the mobile gateway is connected to the local network. It has three functionalities: the first is to discover DLNA-conformant devices in the local network. The second is to send UPnP messages decapsulated from SOAP messages to DLNA-conformant devices. The last is to receive response messages from the local devices and hand them over to the UPnP message relay functionality.

4.2.3. Functionality of the Mobile Gateway

Let us describe the functionality of the mobile gateway

4.2.3.1. UPnP Messaging

UPnP messaging, a functionality that attacks issues 1) and 2) in Section 2.2, is one that sends UPnP messages to the home gateway. Users manipulate the on-screen UI on the mobile device and choose which features of the device they want to activate. The mobile gateway then encapsulates the corresponding UPnP message in SOAP messages and sends it to the home gateway. The home gateway reacts as described in Section 4.2.2.

4.2.3.2. Port Opening Request

This is a functionality that addresses issue 3) described in Section 2.2. The mobile gateway first encapsulates an AddPortMapping UPnP message in a SOAP message and sends it to the home gateway. The home gateway, upon receiving this SOAP message, decapsulates the original UPnP message, and goes on to execute the actions described in Section 4.2.2.

4.2.3.3. Description Retrieval

This functionality checks the DLNA-conformant devices on the home network and their available features. The mobile gateway sends an HTTP GET request to the home gateway to retrieve the Device Descriptions and Service Descriptions of the DLNA-conformant devices.

4.2.3.4. Digital Content Transfer

This functionality lets users transfer digital contents from home DMSs to remote DMPs or DMRs without passing it through the mobile gateway. The mobile gateway retrieves the content list and URIs from the home DMS. It then rewrites the local IP addresses contained in it to global IP addresses. **Figure 2** shows an example content URI before rewriting, and **Figure 3** shows it after rewriting. The user then manipulates the on-screen UI on the mobile device and selects a content to play back. Finally, the mobile gateway sends a SetATransportURI message to the DMP or DMR with the rewritten URI of the selected content.

```
<DIDL-Lite xmlns="urn:schemas-upnp-org:metadata-1-0/DIDL-Lite/"
xmlns:dc="http://purl.org/dc/elements/1.1/" xmlns:upnp="urn:schemas-upnp-org:metadata-1-0/upnp/">
 <item id="0000000000000006" parentID="0000000000000005" restricted="1">
  <dc:title>Animal</dc:title>
  <upnp:class>object.item.videoItem</upnp:class>
  <res protocolInfo="http-get:*:video/x-ms-wmv:*">http://192.168.24.35:57491/Animal.wmv</res>
 </item>
 <item id="0000000000000007" parentID="0000000000000005" restricted="1">
  <dc:title>desktop</dc:title>
  <upnp:class>object.item.imageItem.photo</upnp:class>
  <res protocolInfo="http-get:*:image/jpeg:*">http://192.168.24.35:57491/Tulip.jpg</res>
 </item>
</DIDL-Lite>
```

Figure 2. Contents URI before rewriting.

```
<DIDL-Lite xmlns="urn:schemas-upnp-org:metadata-1-0/DIDL-Lite/"
xmlns:dc="http://purl.org/dc/elements/1.1/" xmlns:upnp="urn:schemas-upnp-org:metadata-1-0/upnp/">
 <item id="0000000000000006" parentID="0000000000000005" restricted="1">
  <dc:title>Animal</dc:title>
  <upnp:class>object.item.videoItem</upnp:class>
  <res protocolInfo="http-get:*:video/x-ms-wmv:*">http://123.227.77.41:57491/Animal.wmv</res>
 </item>
 <item id="0000000000000007" parentID="0000000000000005" restricted="1">
  <dc:title>desktop</dc:title>
  <upnp:class>object.item.imageItem.photo</upnp:class>
  <res protocolInfo="http-get:*:image/jpeg:">http://123.227.77.41:57491/Tulip.jpg</res>
 </item>
</DIDL-Lite>
```

Figure 3. Contents URI after rewriting.

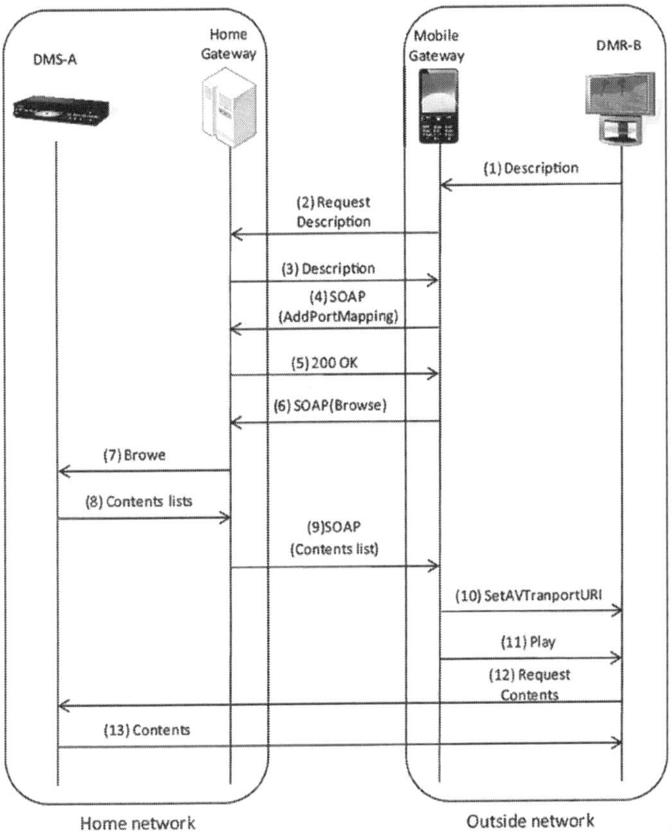

Figure 4. Sequence of proposed method.

4.2.3.5. Controller

The controller makes it possible to manipulate devices on both the home network and the remote network as if they all reside on a single network. The mobile gateway uses UPnP messaging as described above to manipulate devices on the home network. It also implements DMC functionalities as defined in the DLNA Guidelines to manipulate devices on the remote network.

4.3. Example Action Sequence

Let us show an example of the action sequence that takes place when playing back digital contents placed on a home DMS at a remote DMR. **Figure 4** is the sequence diagram. Here, DMS-A is the DMS residing at home and DMR-B is the DMR residing on the remote network.

1) Upon connecting to a network, the mobile gateway uses its controller functionality to retrieve DMR-B's Device Description and Service Description. This lets the mobile gateway manipulate DMR-B.

2) The mobile gateway uses its Description retrieval functionality to send an HTTP GET message and retrieve DMS-A's Device Description and Service Description.

3) The home gateway, upon receiving the HTTP GET message, replies with DMS-A's Device Description and Service Description.

4) The mobile gateway uses its port opening request functionality to request that the port corresponding to DMS-A be opened. More specifically, it sends an AddPortMapping message encapsulated in a SOAP message to the home gateway. The home gateway decapsulates it and sends the AddPortMapping message to the router.

5) The router opens the port described in the AddPortMapping message. A 200 OK message is returned to the home gateway if the port is successfully opened. The home gateway relays this response to the mobile gateway.

6) The mobile gateway, in order to retrieve the contents list and URIs, uses its UPnP messaging functionality to send a Browse message to the home gateway. Again, this message is encapsulated in a SOAP message.

7) The home gateway uses its relay functionality to decapsulate the Browse message and sends it to DMS-A.

8) DMS-A replies to the home gateway with the list of digital contents and URIs.

9) The home gateway encapsulates the list in a SOAP message and sends it to the mobile gateway.

10) The mobile gateway uses its digital content transfer functionality and rewrites local IP addresses contained in the list to global IP addresses. The user then selects which content to play back. The mobile gateway then generates a SetAVTransportURI message with the URI of the selected item and sends it to DMR-B. DMR-B then waits for a Play message.

11) The mobile gateway sends a Play message to DMR-B.

12) DMR-B, upon receiving the Play message, sends an HTTP GET message to DMS-A based on the URI that was set with SetAVTransportURI.

13) DMS-A sends the requested digital content to DMR-B. Note that the content is sent directly without passing through the mobile gateway.

5. EVALUATION

5.1. Evaluation Items and Environment

We compared our proposed method with a regular local DLNA setup and evaluated its practicality. We evaluated the following items.

1) Elapsed time before the list of digital contents was received after requesting it to the DMS.

2) Elapsed time before the digital content playback started after requesting it to the DMR.

3) Number of times the content playback was interrupted for buffering.

In item 1), we measured the elapsed time before the content list is retrieved after the mobile gateway sends the Browse message to the DMS. In item 2), we compared two values: the elapsed time before playback begins after a local DMC sends the Play message to the local DMR, and the elapsed time before playback begins after the mobile gateway sends the Play message to the remote DMR. We measured 1) and 2) 20 times each with the packet-capture application WireShark and took the average. In item 3), we counted the number of times music or video playback was interrupted for buffering. We did this 20 times and took the average as well.

The following are the implementation environment and the contents used. **Table 1" target="_self"> Table 1** shows the environment and devices used in evaluation. The digital contents used are as follows: JPEG image files in sizes 25 KB, 100 KB and 500 KB each, MP3 music files in bitrates 64 Kbps, 128 Kbps and 224 Kbps each, WMV video files in 688 Kbps, 2.1 Mbps and 6.1 Mbps each. We used these three files for each format to evaluate the setup with low, medium and high quality contents. The lengths of the music files were 3 minutes 15 seconds for 64 Kbps, 4 minutes 23 seconds for 128 Kbps and 3 minutes 09 seconds for 224 Kbps. The lengths of the video files were 3 minutes

30 seconds for 688 Kbps, 3 minutes 30 seconds for 2.1 Mbps and 3 minutes 38 seconds for 6.1 Mbps.

The network architecture for the regular local DLNA setup is shown in **Figure 5" target="_self">Figure 5**. The network architecture for our proposed method is shown in **Figure 6**. The home network for the proposed method was a 100 Mbps FTTH connection provided by OCN Hikari with Flet's. The remote network for the proposed method was Science Information Network's SuperSINET3.

5.2. Results and Discussions

Evaluation results were as follows. **Figure 7** shows results for evaluation item 1). The latency of content list retrieval for the proposed method was 95 msec longer than that for the regular DLNA setup. This difference is barely perceptible, so the mobile gateway is practical for manipulating DLNA-conformant devices on the home network.

Table 1. Comparative environment.

HGW	OS	Windows XP professional SP3
	CPU	Intel Core 2 Quad (2.33 GHz)
	RAM	2 GByte
	Developing environment	Cyber link for Java
MGW	OS	Windows XP professional SP3 Cel
	CPU	eron U2300 (1.20 GHz)
	RAM	1 GByte
	Developing environment	Cyber link for Java
DMS	Intel AV media server	
DMC	Intel AV media controller	
DMR	Intel AV media renderer	

Figures 8-10 show results for evaluation item 2). We observed longer playback latencies for all three kinds of digital contents used. This is due to the slow connection of the Internet compared to the local network.

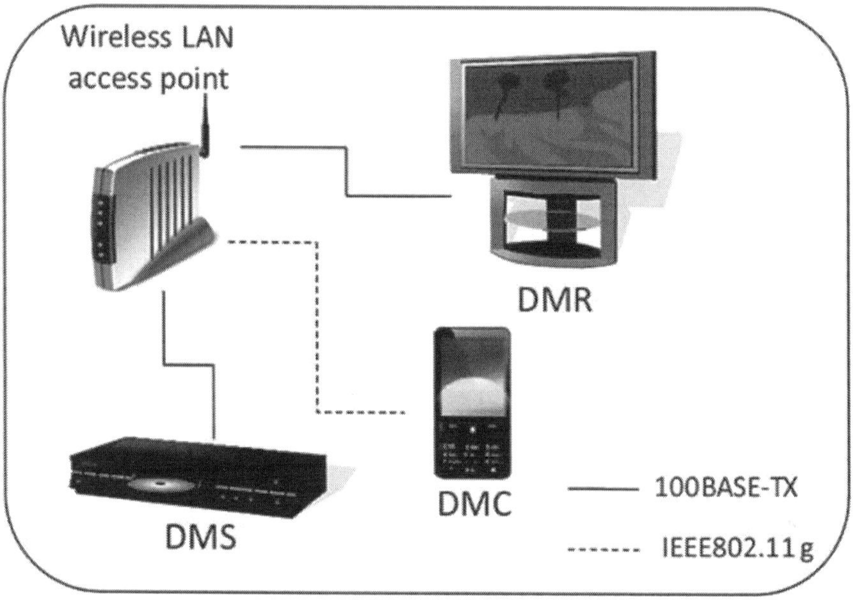

Figure 5. Evaluate network of standard DLNA.

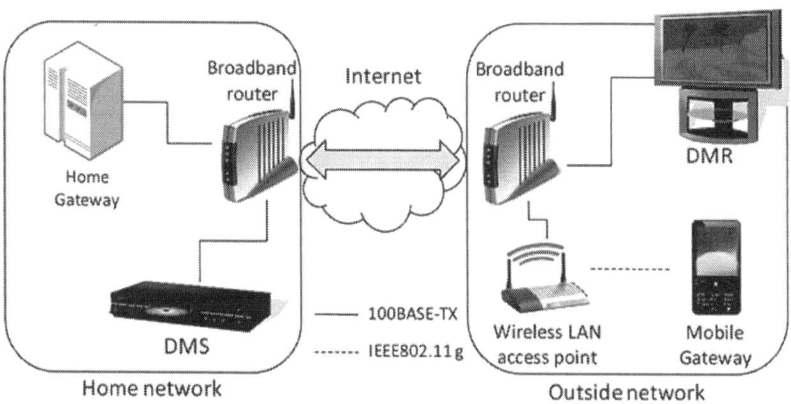

Figure 6. Evaluate network of proposed method.

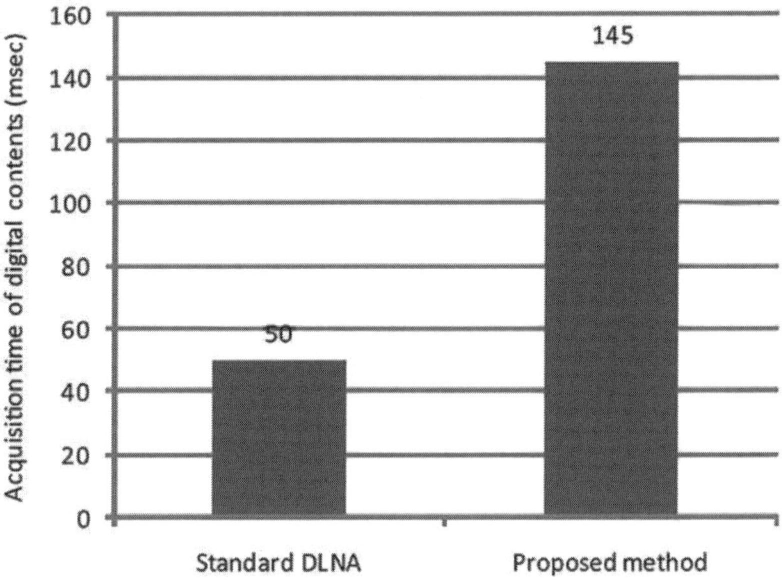

Figure 7. Acquisition time of digital contents.

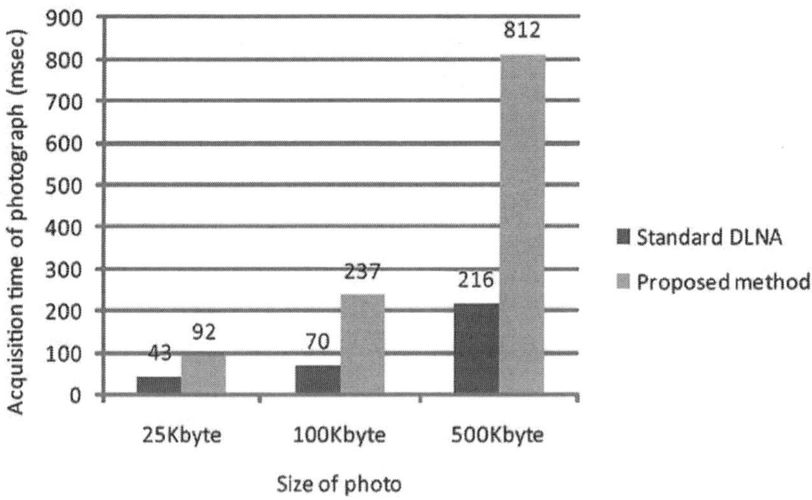

Figure 8. Acquisition time of photograph.

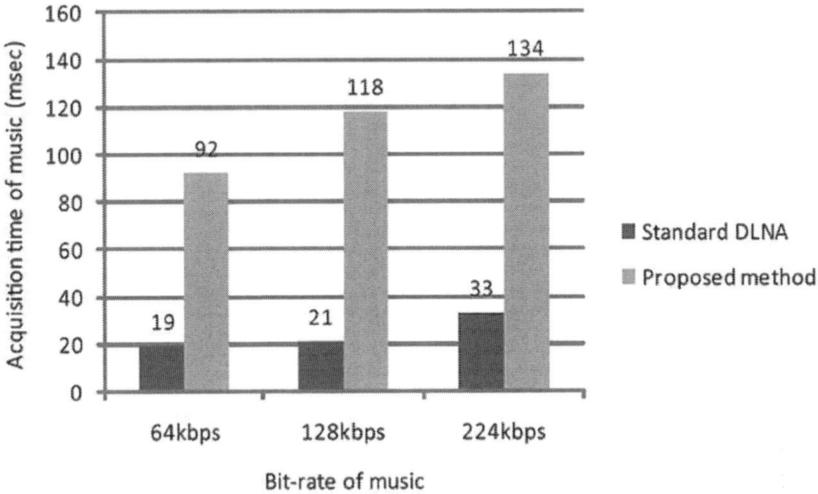

Figure 9. Acquisition time of music.

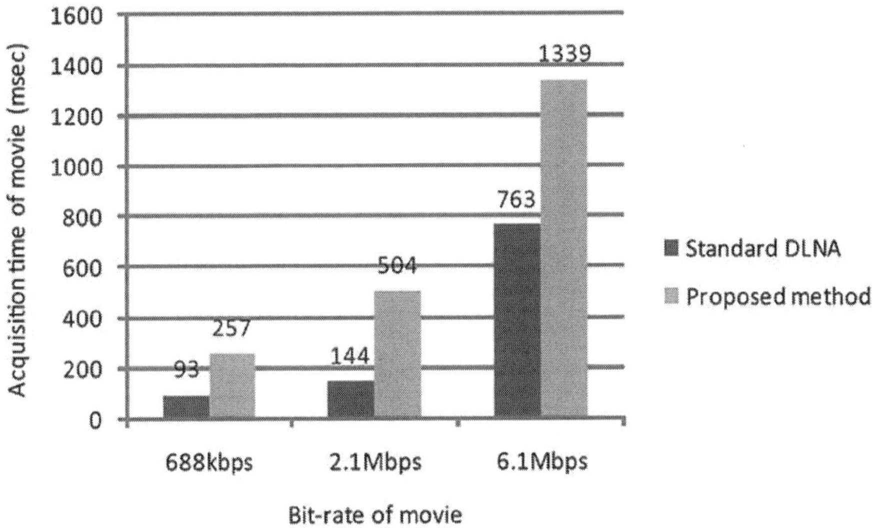

Figure 10. Acquisition time of movie.

Table 2. Interruption frequency of digital contents.

Kind of content	Music			Movie		
Bit-rate (kbs)	64	128	224	688	2100	6100
Thodem Proposed (times)	0.0	0.0	0.0	0.0	0.0	1.3
Standard DLNA (imest)	0.0	0.0	0.0	0.0	0.0	0.4

Table 2 shows results for evaluation item 3). We experienced buffering interruptions of approximately 2 seconds in length when playing back the 6.1 Mbps video, both with the proposed method and the regular DLNA setup. We also confirmed that the other two videos of lower quality played back without interruptions, again both with the proposed method and the regular DLNA setup. The proposed method is practical as video can be played smoothly as long as the bit rate is lower than the Internet connection speed.

6. COMPARISONS WITH RELATED RESEARCHES

6.1. Mobile Wormhole Device

The Mobile Wormhole Device [14] places the Wormhole Device described in Section 3.1.2 on a mobile phone. When the phone is connected to a public wireless LAN access point, it can connect to the other Wormhole Device residing at home and achieve Wide Area DLNA Connectivity. The Mobile Wormhole Device lets users play back digital contents at remote locations even if a Wormhole Device is not (or cannot be) set up.

6.2. W-DLNA

With W-DLNA (Wide Area Digital Living Network Alliance) [15,16], gateways are placed at the borders of the local networks. The gateway at home generates a virtual delegate DMS that acts in place of the DMS in the remote network. Similarly, the gateway at the remote network generates a virtual delegate DMP that acts in place of the DMP in the home network. These gateways use SIP messages to relay messages among DLNA-conformant devices, thus achieving Wide Area DLNA Connectivity.

6.3. Mobile Home to Home

Mobile Home to Home [17] uses wireless LAN-enabled mobile devices to achieve Wide Area DLNA Connectivity. Because it uses mobile devices to enable playback of digital contents on home DMSs at remote DMPs, no additional software or hardware needs to be installed at remote networks.

6.4. Comparison

Table 3 is a comparison of functionalities among these methods.

Regarding the environments in which Wide Area DLNA Connectivity is made available, our proposed method, the Mobile Wormhole Device and Mobile Home to Home use wireless LAN-enabled mobile devices and W-DLNA uses mobile phone lines. Thus, all methods achieve Wide Area DLNA Connectivity without the need to install additional gateways.

Table 3. Comparison with related research.

	Proposal	MWD	W-DLNA	MH2H
Environment	OK	OK	OK	OK
Global connection	OK	NG	OK	OK
GW load reduction	OK	NA	NG	OK
Operation	OK	OK	OK	NG
Version	1.5	1.0	1.0	1.5

MWD: Mobile Worm Hole Device; MH2H: Mobile Home to Home.

Regarding Wide Area DLNA Connectivity, the mobile devices used with the Mobile Wormhole method does not support the manipulation of DMPs and DMRs placed away from home. Thus, it cannot play back contents on the home DMS at remote DMPs and DMRs. In contrast, our proposed method implements digital content transfer and controller functionalities, so it can manipulate remote DMPs and DMRs to play back digital contents.

Regarding processing load at the gateways, W-DLNA transfers digital contents to DMPs via gateways, so processing load is heavy. In contrast, our proposed method uses its digital content transfer functionality to transfer contents without passing it through the mobile gateway, so processing load is trivial.

Regarding limitations in device manipulation, Mobile Home to Home does not manipulate devices via a home gateway but interacts with them directly. This introduces risks of malicious operations on digital contents such as unauthorized deletions. In contrast, our proposed method manipulates devices via a home gateway, so administrators can limit operations on the contents.

Regarding DLNA Guideline versions, W-DLNA does not support DLNA Guidelines version 1.5. In contrast, our proposed method implements UPnP messaging and controller functionalities in the mobile gateway, so it can communicate with devices supporting DLNA Guidelines version 1.5.

7. CONCLUSION

Existing researches required home gateways to be installed in remote networks to achieve Wide Area DLNA Connectivity, thus limiting the environments. This paper proposed a method that expanded on our SOAP method to use mobile devices such as notebook PCs, mobile phones and smartphones as mobile gateways and achieve Wide Area DLNA Connectivity. With this method, Wide Area DLNA Connectivity can be achieved even in environments where home gateways are not (or cannot be) installed. The proposed method also makes it possible to manipulate devices conforming to DLNA Guidelines version 1.5. Moreover, because the mobile gateway transfers contents from home DMSs to remote DMPs and DMRs directly, processing load at the mobile gateway is kept low. We evaluated its performance in terms of the latency of content list retrieval, the latency of playback, and buffering interruptions during playback. All of these metrics proved to be practical. In future researches we plan to evaluate the method in terms of overhead introduced by encrypting data transfers between the home gateway and the mobile gateway.

8. ACKNOWLEDGEMENTS

This work was supported by JSPS Grant-in-Aid for Scientific Research KAKENHI (2100084).

REFERENCES

1. Digital Living Network Alliance, "DLNA Networked Device Interoperability Guidelines Volume 1: Architecture and Protocols," 2006.
2. UPnP Forum, UPnP AV Architecture Version 1.0, 2002.
3. K. W. Kim, Y.-J. Lee, T.-B. Lim and S.-P. Lee, "Design and Implementation of Aggregation Framework for UPnP/DLNA Media Server," Proceeding of 20th International Conference on Computer Communications and Networks, Maui, 31 July 2011, pp. 1-4.

4. T. Hwang, H. Park, E. Paik and J. Chung, "EAFR-Based DLNA Proxy for High-Quality Video Distribution in Extended Home Space," IEEE Transactions on Consumer Electronics, Vol. 57, No. 1, 2011, pp. 120-125.
5. M. Ogawa, H. Hayakawa, T. Koita and K. Sato, "Transparent UPnP Interactions over Global Network," Proceedings of International Conference on Mechatronics and Information Technology, (ICMIT 2007), Gifu, 5 December 2007, 67944 p.
6. K. Nakamura, M. Ogawa, T. Koita and K. Sato, "Implementation and Evaluation of Caching Method to Increase the Speed of UPnP Gateway," Proceedings of the 2008 IEEE/IFIP International Conference on Embedded and Ubiquitous Computing, Shanghai, 20 December 2008, pp. 112-118.
7. World Wide Web Consortium, "W3C Recommendation (Second Edition), SOAP Version 1.2 Part 0: Primer," 2007. http://www.w3.org/TR/2007/ REC-soap12-part0-20070427/
8. World Wide Web Consortium, "W3C Recommendation (Second Edition), SOAP Version 1.2 Part 1: Messaging Framework," 2007. http://www.w3.org/TR/2007/REC-soap12-part1-20070427/
9. World Wide Web Consortium, "W3C Recommendation (Second Edition), SOAP Version 1.2 Part 2: Adjunct," 2007. http://www.w3.org/TR/2007/ REC-soap12-part2-20070427/
10. D. Muto and T. Yoshinaga, "Wormhole Device: Software-Assisted Remote Communication Mechanism for DLNA-Based Appliances," Proceedings of Multimedia, Distributed, Cooperative, and Mobile Symposium, Mie, 4 July 2007, pp. 134-138.
11. J.-Y. Wu, D. Muto, T. Koyama and T. Yoshinaga, "Mechanism for Sharing Media Content in Multiple Home Network Environments," IEICE Technical Report, Vol. 108, No. 138, 2008, pp. 93-98.
12. T. Haruyama, S. Mizuno, M. Kawashima and O. Mizuno, "Dial-to-Connect VPN System for Remote DLNA Communication," Proceedings of IEEE Consumer Communications and Networking Conference, Las Vegas, 10-12 January 2008, pp. 1224-1225.
13. Y.-J. Oh, H.-K. Lee, E.-H. Paik and K.-R. Park, "Implementation of the DLNA Proxy System for Sharing Home Media Contents," IEEE Transaction on Consumer Electronics, Vol. 53, No. 1, 2007, pp. 139-144.

14. T. Koyama, J.-Y. Wu, D. Muto and T. Yoshinaga, "Mobile-Wormhole Device: An Application that Implements a Mechanism Called Assisted Remote Communication for a Mobile Device," IPSJ SIG Technical Report, Vol. 2008-UBI-017, No. 18, 2008, pp. 1-8.
15. S. Motegi, K. Tasaka, A. Inoude and H. Horiuchi, "Proposal on Wide Area DLNA Communication System," Proceedings of IEEE Consumer Communications and Networking Conference, Las Vegas, 10-12 January 2008, pp. 233-237.
16. S. Motegi, K. Tasaka, N. Tehpvilojanapong and H. Horiuchi, "Proposal on Wide area DLNA Communication System," IEICE Technical Report, Vol. 107, No. 6, 2007, pp. 71-76.
17. M. Miyake, T. Yoshikawa, S. Ookubo and A. Takeshita, "Mobile Home to Home: Remote Access for DLNA Device between Home Networks," IPSJ SIG Technical Report, Vol. 2008-MBL-47, No. 107, 2008, pp. 9-15.

Chapter 12

Securing Health Sensing Using Integrated Circuit Metric

Ruhma Tahir *, Hasan Tahir and Klaus McDonald-Maier

Embedded and Intelligent Systems Research Laboratory, School of Computer Science and Electronic Engineering, University of Essex, Wivenhoe Park, Colchester CO4 3SQ, UK

ABSTRACT

Convergence of technologies from several domains of computing and healthcare have aided in the creation of devices that can help health professionals in monitoring their patients remotely. An increase in networked healthcare devices has resulted in incidents related to data theft, medical identity theft and insurance fraud. In this paper, we discuss the design and implementation of a secure lightweight wearable health sensing system. The proposed system is based on an emerging security technology called Integrated Circuit Metric (ICMetric) that extracts the inherent features of a device to generate a unique device identification. In this paper, we provide details of how the physical characteristics of a health sensor can be used for the generation of hardware "fingerprints". The obtained fingerprints are used to deliver security services like authentication, confidentiality, secure admission and symmetric key generation. The generated symmetric key is used to securely communicate the health records and data of the patient. Based on experimental results and the security analysis of the proposed scheme, it is apparent that the proposed system enables high levels of security for health monitoring in resource optimized manner.

Keywords: ICMetric; accelerometer; authentication; confidentiality; remote healthcare; secure remote password protocol

1. INTRODUCTION

Healthcare devices are now increasingly being networked, thus also raising concerns about their security and safety. Attackers attempt to penetrate health monitoring devices with an intention of stealing data so that it can be used for a range of purposes ranging from medical identity theft to prescription theft. Health monitoring devices are available for a variety of purposes like medical, fitness, entertainment and lifestyle monitoring often described as "quantified self" [1]. These can be both invasive and noninvasive sensors/devices. Devices like pacemakers can be implanted into a patient's body while smart watches and wearable sensors function externally. When these healthcare devices become part of a network like the Internet of Things, we are faced with a situation where the devices are exposed to potential attacks. There are sensors that monitor a person's vital signs like breathing, heart rate and body temperature at regular intervals. Since these sensors monitor a person's physiological responses, it is imperative to secure the device, data, patient and health professional in a resource efficient but comprehensive manner [2].

Traditionally, cryptographic algorithms provide security in various applications by using stored encryption/decryption keys. These algorithms have the inherent disadvantage that, if the device is compromised, the keys could be revealed to the adversaries, which would result in important data being exposed. We propose the use of Integrated Circuit Metric (ICMetric) [3] as an alternative to stored keys [4].

An advantage of the scheme presented here, is that human intervention is not required at any point during the lifecycle of its secure functioning. This is particularly important from the patient's perspective, since this will ensure its usefulness while securely transmitting health data. Thus, the security functions of our framework are based on ICMetric data derived from the respective devices. The proposed health sensing system has the following features, which address the need for secure remote monitoring of patients:

- In this system the ICMetric technology is utilized to generate keys based on the hardware/software characteristics and specification of the device, thereby enabling device verification. This provides an effective means to address the issues related to storage of key, thereby safeguarding against the major threat of key compromise. In our work we highlight how such a device identification can be generated for a health sensor using its physical characteristics, e.g., from the Micro Electro Mechanical Systems (MEMS) accelerometer. In addition to the accelerometer credentials, we identify a range of other features that also form part of the device identification.

- In this system we propose a secure admission control process as a core component. This part of the design is aimed at recognizing each entity, which is part of the application. It recognizes each patient based on its allocated sensors, and the health professionals based on their end point device. So this part authenticates, authorizes and evaluates each entity prior to secure communication of the health data. However, at no point is the ICMetric value sent over the channel or even communicated to others.
- A challenge with the generated ICMetric is the entropy and length of the generated secret value. The proposed design generates a strong symmetric key based on the communicating parties' device ICMetric. This generates symmetric key facilitates in secure end-to-end communications.

The remainder of this paper is organized as follows: Section 2 discusses in detail the ICMetric technology and its design principles. Section 3 provides a brief discussion on MEMS and we highlight the structure of a MEMS accelerometer as an example. This section provides details of how an ICMetric identification can be generated using MEMS accelerometers and other features specific to a health sensor. Section 4 presents the basic system design along with the design goals. The detailed design of our proposed framework for the generation of strong high entropy keys is presented in Section 5. Implementation details of the design are discussed in Section 6.Section 7 offers an in-depth analysis of the scheme from various perspectives like security, scalability and stability. This section also presents a detailed security analysis of the scheme. The paper concludes in Section 8 by summarizing the findings of this work.

1.1. Related Work

In [5], authors present a comprehensive survey on various aspects of Wireless Body Area Networks. In their paper, they have discussed various application scenarios, their context and specific requirements related to the technology.

Bourbakis *et al.* in [6] have proposed a mobile health platform for protecting health data during exchange with a server. This research includes various biometric authentication systems to incorporate authentication, authorization, confidentiality and integrity services in their secure symmetric health monitoring systems.

In [7], the authors proposed a scheme called Cryptographically enforced and Privacy Enhanced (CRYPE) for secure one to one communications. It uses identity based encryption for secure end to end communication and provides confidentiality and role based access control.

Yi *et al.* [8] have proposed an approach to secure communication between the sensor node and the administrator using a lightweight encryption algorithm. Furthermore, their design tries to safeguard from inside attacks by making use of the "Sharemind" system. In their proposed scheme, the authenticate-reject decision for a specific sensor is based on the outcome of three data.

Barua *et al.* [9] propose a patient centric access control scheme that guarantees integrity and confidentiality of patients health data by using digital signature and pseudo-identity techniques.

Several researchers are also looking into the design and development of key generation and key agreement schemes in BANs. The use of physiological data coupled with cryptographic schemes are two principal preferences in line with this work [10,11,12,13,14].

The development in networked health monitoring systems has led to research being carried out on resource efficient cryptographic protocols. Researchers are now decoupling the medical and security functionalities to propose resource efficient architectures executing on two separate processing elements/cores [15]. For implementing the proposed security framework; we have used AES-NI for the very purpose, to provide a lightweight scheme for remote secure health monitoring. Other works in this line include the SCAN secure processor [16], which supports secure biometric authentication and various symmetric encryption primitives.

Since the ICMetric technology is a relatively newer concept, most work has been done to prove its effectiveness and applicability. Perhaps the most extensive work on ICMetric has been done by Papoutsis. In his work [17] he proves that it is both feasible and suitable to use the hardware features of a system to generate hardware identifications. Recent studies [3] have coupled the ICMetric technology with an intelligent [18] battery powered wheelchair to provide security in health care services.

2. INTEGRATED CIRCUIT METRIC (ICMETRIC)

Security algorithms rely on using stored secret keys. Cryptographers increase the key size to stop keys from being brute forced, in times when computation devices become cheaper and more common. Increasing the key size can deter attackers who are attempting to brute force the cryptographic keys, however this increase in key size cannot deter key theft by infiltration or simple theft. When the keys of a system are exposed, the security of the system is compromised in its entirety.

Using characteristic features specific to a device, an identification can be generated for every computation device. The device identification formally called the ICMetric is generated each time it is required and discarded thereafter thus entirely eliminating infiltration and theft based attacks. What differentiates the concept of ICMetric technology from other hardware fingerprinting techniques is the choice of features that are used for generating device identification. Conventional fingerprinting techniques rely on using features which are static, easy to capture and use. While using features that are easy to extract one is also faced with a problem that these features can be easily replicated/spoofed by attackers, thus potentially rendering the hardware fingerprint useless. Some features that can be used for generating an ICMetric are MAC addresses, CPU IDs and serial numbers. To increase the complexity of the generated ICMetric, other features are employed which are application usage specific like camera resolutions, GPS coordinates, browsing histories, common user files, system profiles *etc.* The advantage of using variable internal features is that they are difficult for an attacker to predict/spoof at runtime.

If the device fingerprint is generated by using features that are easily extractable, then the fingerprint lacks diversity and entropy. For instance even though a MAC address is a unique feature to a device, it can be extracted and hence spoofed. Some features of a device cannot be used because either they cannot be extracted or they do not adequately assist in identifying a device. The generation of the ICMetric is a two phase process, *i.e.*, calibration phase and the operation phase.

2.1. Calibration Phase

The calibration phase is applied once and only when the ICMetric is required. In this phase feature values are extracted from the system and normalized. In this phase the feature maps are studied for their standard deviation, confidence interval, interquartile range and skewness. A multitude of feature values are read to ensure a wide coverage of features.

2.2. Operation Phase

In the operation phase the feature values are combined to form an ICMetric number. Two techniques can be used for combining the individual feature values, *i.e.*, feature addition and feature concatenation [17]. In the feature addition technique the individual feature values are added to produce a shorter but highly diverse device ICMetric. The feature concatenation technique produces an ICMetric by concatenation of the individual feature values. The produced ICMetric is longer in length but lacks diversity as compared to the ICMetric generated by using feature addition technique.

2.3. Generating the Device ICMetric

To deter key and identification theft, the ICMetric of a device is not stored or transmitted. The ICMetric will be generated when required, therefore its generation must be comprised of lightweight mathematical functions. Generation of the ICMetric is a process requiring mathematical and statistical analysis of feature values. If a feature has a single static value that does not change, the feature does not need any form of manipulations. Whereas if the feature exhibits a range of values then the feature values need to be analyzed and studied statistically. When using the ICMetric technology the focus is on using features that have a definite range. This means that the feature values must be present in the form of clusters so that they can be placed on a normal distribution. Once this is achieved the normal distributions are analyzed, so that meaningful values can be extracted from graphical representations.

3. TYPES OF HEALTH SENSING DEVICES

Recent developments in various domains like telecommunication, microelectronics and health sensing has resulted in the emergence of small scale wearable devices which can detect human physiological signals with minimum manual intervention. Due to the small size of sensors it is now possible to embed them into everyday objects like clothes, wearable accessories *etc.* Health sensors can be placed into many individual categories, however it is appropriate to classify sensors based on their purpose/aspects:

- Environment—sensors designed to detect temperature, humidity, pressure, light, noise *etc.*
- Device—some sensors are designed to sense physiological data directly from the human body, while other sensors use images like facial features, eye movement to measure data.
- User—sensors designed to detect user specific data like location, acceleration, heart rates, EMI, body temperature, *etc.*
- Interaction—sensors designed to interact with multiple device through diverse mediums like RFID, Bluetooth, Ethernet *etc.* Pervasive sensors allow multiple sensors to interact with each other.

Modern devices are fundamentally a collection of many individual sensors that operate in harmony. This collection of interoperating sensors allows devices to provide extended range of services.

3.1. MEMS Accelerometer

MEMS is a technology that was first studied [19] in 1986 and it is now utilized in a wide range of devices [20]. MEMS technology works at micro scale to create efficient electrical systems that are based on mechanical elements. If two sensors are given the same stimuli then we can say with a high level of certainty that both sensors will provide different results [21]. This is precisely why sensor based technologies need calibrations and adjustments. This variance is due to how MEMS are manufactured. Minor imperfections in the fabrication process can have significant impact on how the sensors behave. The MEMS accelerometer is highly sensitive and imperfection prone sensor which is adversely effected by the fabrication process. Slight stresses introduce a permanent offset which varies from sensor to sensor and cannot be predicted. The use of this offset is sufficient for the establishment of a device authentication. Recent research [22,23,24,25,26] in the field of sensor based device identification proves that it is both possible and feasible to generate device identifications by using sensory data. Practically it is not always easy/feasible to extract a device imperfection. For instance to measure the offset of gyroscopes, a special device is required that can rotate the target device at a constant angular velocity. Construction of such a device is a complex task in the lab and the same behavior cannot be replicated while the sensor is put into practical use. Given in Table 1 are various sensors and their bias which can be used to form an identification.

Table 1. Sensors and their associated bias.

Sensor	Bias
Accelerometer	Linear acceleration bias
Gyroscope	Linear gyroscopic bias
Touchscreen	Screen misalignment
Camera	Camera noise pattern
GPS	Clock skew

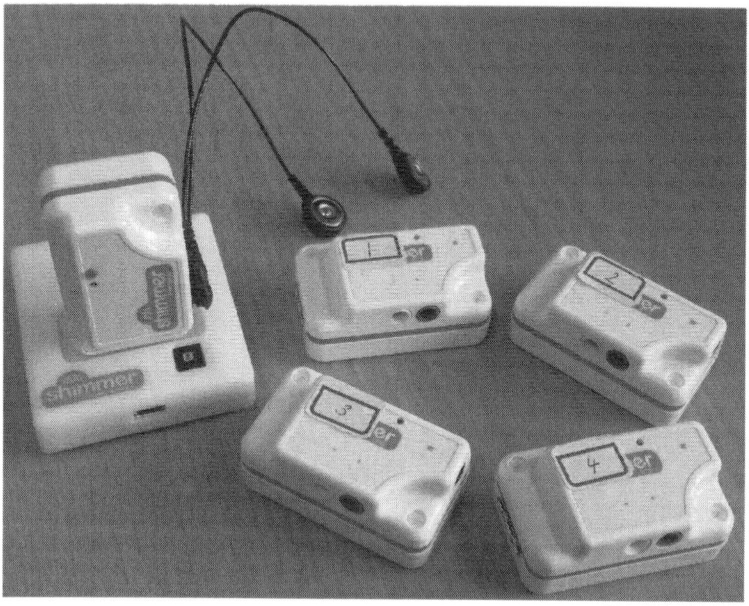

Figure 1. A collection of five Shimmer sensors used to determine the accelerometer bias.

3.2. Accelerometer Offset Measurement and Device Feature Establishment

To determine the offset of the accelerometer the sensing device must be placed on a stable surface which is free from vibrations and erratic movements. Using the Shimmer [26] as an experimental platform we placed the sensors on such a surface and obtained 1552 individual readings from the accelerometer sensor. The Shimmer sensor is embedded with a Freescale MMA 7361 [27] that has two selectable sensitivities, *i.e.*, ±1.5 g and ±6 g. The output CSV file is composed of a number of values that are read at regular intervals. The data provided includes a raw timestamp, calibrated timestamp (microseconds), raw accelerometer values, calibrated accelerometer reading (m/s^2). Figure 1 shows the experimental setup composed of five individual sensors that transmit readings via Bluetooth. One sensor is plugged into the docking base station.

As highlighted above, the accelerometer is becoming a standard sensor in many computation devices. Hence, the proposed scheme for the extraction of an accelerometer offset can be used in any computation device that possesses an accelerometer sensor.

A candidate feature for the generation of an ICMetric must possess the property that its data can be normalized. This is a necessary requirement because if the data cannot be normalized, then the data cannot possess a deterministic range. Analysis of the normal distributions comprises of determining the confidence interval, standard deviation, interquartile range, variance, mean value and the skewness of the normal distribution. The confidence interval gives the guarantee that the mean of an interval is contained within the defined limits of the normal distribution. The confidence interval is an important characteristic, because if the accelerometer exhibits erratic behaviour then the mean of an individual axes will also be displaced resulting in a displaced confidence interval. To generate the device ICMetric, we have used a 95% confidence interval in order to obtain a high level of certainty. If $\bar{X}$ is the mean of the axis, n is the number of individual values in the distribution and σ is the standard deviation then the confidence interval CI is given by:

$$95\%\ CI = \bar{X} \pm 1.96\frac{\sigma}{\sqrt{n}} \tag{1}$$

The CI is composed of a numeric multiplier of 1.96. This multiplier gives a two sided confidence interval of 95%.

Another statistical indicator which is used in the analysis of the normal distribution is the probability mass function. The probability mass function is a function f(x), which defines the probability that the random variable takes a particular value in a fixed range thus,

$$f(x) = \frac{1}{\sigma\sqrt{2\pi}} e^{\frac{-(x-\mu)^2}{2\sigma^2}} \quad (2)$$

where σ2 is the variance.

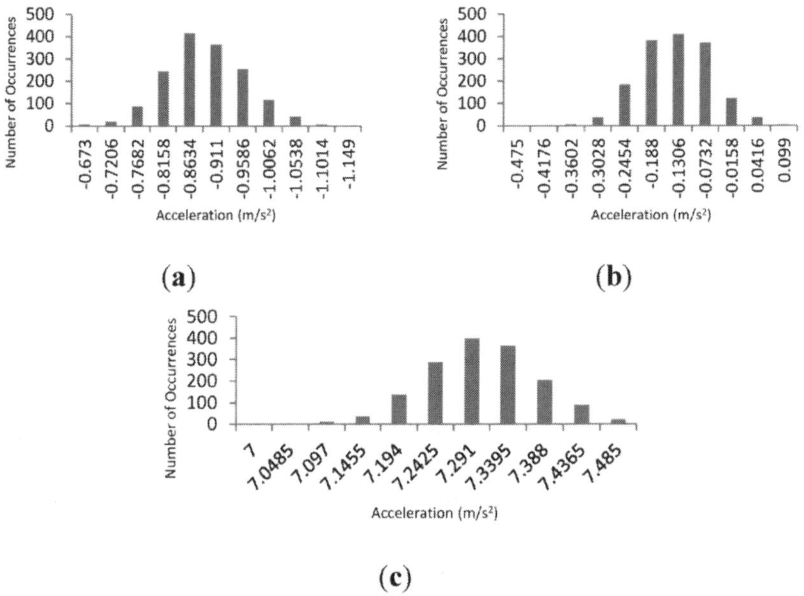

Figure 2. Calibrated accelerometer normalization graph for the (**a**) *x*-axis; (**b**)*y*-axis; and (**c**) *z*-axis.

The normal distribution of the readings obtained from the *x*, *y* and *z* axis show that the accelerometer readings can be normalized and used for the establishment of an ICMetric number. Statistical and mathematical analysis of the normal distributions shown in Figure 2 proves that distinct values accurately reflect the accelerometers responses along the three axes. Table 2 provides the analysis of the normal distribution for two different devices. The values show that the bias can differ dramatically, which makes it a strong metric for establishing the device ICMetric.

Table 2. Statistical analysis of the normal distribution for two different devices A and B.

		Device A	Device B
x-axis	Confidence Interval	(−0.919,−0.912)	(−0.281,−0.298)
	Standard Deviation	0.071	0.0693
	Inter Quartile Range	−0.285	−0.473
	Mean	−0.916	−0.292
	Skewness	−1.190	0.0229
y-axis	Confidence Interval	(−0.160,−0.167)	(−9.248,−9.271)
	Standard Deviation	0.074	0.0678
	Inter Quartile Range	0.344	0.411
	Mean	−0.163	−9.266
	Skewness	0.072	0.0685
z-axis	Confidence Interval	(7.284,7.276)	(−0.598,−0.621)
	Standard Deviation	0.075	0.053
	Inter Quartile Range	0.291	0.0728
	Mean	7.280	−0.615
	Skewness	−31.341	−1.371
	Resulting ICMetric	**−13.286**	**−31.5687**

It is this resulting statistical data which is used to form the final device ICMetric. The values are combined to generate a final ICMetric using either the feature combination or the feature addition technique discussed inSection 2.2.

The ICMetric cannot be generated by just using the bias in an accelerometer. More features can be incorporated so that the generated identifier is truly diverse and ensures that the ICMetric cannot be easily predicted/spoofed. A range of features were identified, which can assist in the formation of the device ICMetric. Although these features have been studied in the Shimmer sensor, the same features can be found in other modern body sensors. Therefore the proposed features are not limited to the Shimmer sensor only. Given below is a range of features that are used for generating the ICMetric:

- Sensor Bluetooth MAC address—a 48 bit MAC address that uniquely identifies each sensor.
- Dongle Bluetooth MAC address—a 48 bit MAC address that uniquely identifies the dongle to which the sensor is connecting.
- Bluetooth radio identification—a 16 bit hexadecimal number that is associated with a device. This identification can be modified which makes it a strong ICMetric candidate.

- Factory lasered 64 bit registration—Every sensor is equipped with a DS-2411 chip [28] which provides the device with a unique identification. This is a single line chip which is factory coded with a serial that can be read for the purpose of identification. The serial number is composed of a CRC, serialization and a family code. The serialization structure is given in the Figure 3:
- Calibration Matrices—it is the role of every sensor to generate readings that are accurate. The calibration is defined by three individual matrices namely the offset vector, sensitivity matrix and alignment matrix.Figure 4 shows a sample set of calibration matrices which can be used to identify a sensor. Here it must be pointed out that the smallest change in the calibration matrices causes a chain of events which results in the wrong ICMetric being generated. When an attacker attempts to guess the calibration matrices, the resulting accelerometer readings will not reflect what was expected from the sensor. This will ultimately result in the wrong ICMetric being generated. The readings are generated following a specific algorithm [29] targeting accelerometer readings and calibrations.

CRC 8 bit	Serialization 48 bit	Family Code 8 bit

Figure 3. The DS-2411 serialization structure.

−1
−1
−1

(a)

65,535	0	0
0	65,535	0
0	0	65,535

(b)

−0.01	−0.01	−0.01
−0.01	−0.01	−0.01
−0.01	−0.01	−0.01

(c)

Figure 4. The calibration matrices with sample values. (**a**) Offset vector; (**b**) Sensitivity Matrix; (**c**) Alignment Matrix.

The experiments using the Shimmer health sensor show that the accelerometer bias is unique and repeatable for each sensor. The *x*, *y* and *z* axis readings of a single sensor show no correlation between each other. A comparative study of our sensors shows that sensors display no particular pattern and that the bias cannot be predicted. The resulting device ICMetric is a singular device identification because the bias is an attribute of the embedded accelerometer

[30]. Combining the accelerometer bias metrics with other inherent device features ensures that the ICMetric is in fact a unique device identification.

4. FRAMEWORK ASSUMPTIONS

Attacks [31,32] on a healthcare system can target the client, KGC or network. Healthcare applications address several security and privacy challenges to secure the healthcare data. Eavesdropping is a major threat to patient privacy; the adversary can easily read unencrypted messages being transmitted over the network. Physiological data can be safeguarded from these attacks by employing stronger security measures in the system that provides privacy to the sensed patient data. The security framework proposed here seeks to improve the security of health monitoring applications by providing secure admission control, authentication, confidentiality and integrity of the health data in combination with the ICMetric technology. The proposed scheme enables networked environments and all entities/devices forming part of the network trust in a Key Generation Center (KGC). The KGC is responsible for controlling individual entities in the network. The KGC enables entities that have never had contact before to interact securely and confidentially. Figure 5 shows the general system model connecting the patient, KGC and the health professional. The KGC is operated by the healthcare provider and is responsible for assigning all the network specific configurations to successfully join a network. All entities have already decided to trust the KGC, either without authentication or on the basis of KGC authentication via SSL.

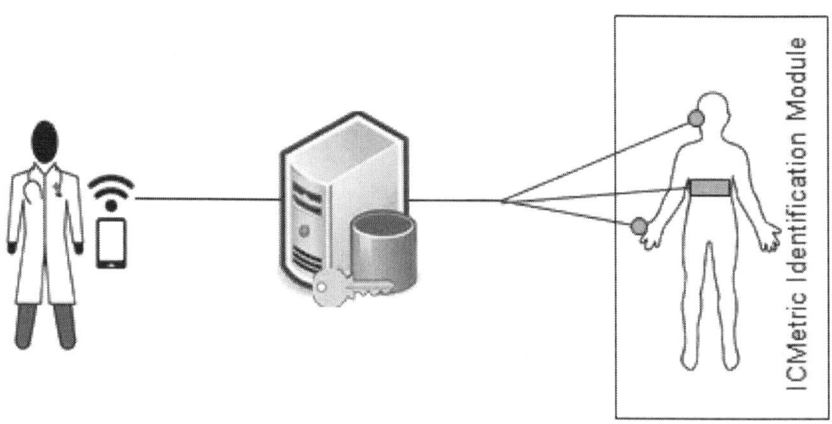

Figure 5. System model for an ICMetric based health monitoring system.

5. THE SECURITY FRAMEWORK

The proposed security framework details a secure health sensing system which combines wearable health monitoring devices with the ICMetric technology to guarantee a system which can deter common attacks on healthcare systems. The following section details the steps involved in the secure functioning of the health monitoring application based on ICMetric:

5.1. Secure Admission Control

The requirement and importance of secure admission control is obvious, as key management and secure communication schemes are effective only after the devices join the network in a secure admission process. The admission control scheme is used only once, when the entity registers for the first time. The registration of an entity is a separate process that requires both manual and electronic data collections. The entity registration consists of the following unique activities:

- When an entity wishes to register with the KGC, its necessary credentials are forwarded to a KGC where the registration is digitized.
- Once the entity is registered, the KGC is updated with a unique identification and a 128 bit random per-entity value (so called "salt").
- A copy of the identification and salt is also provided to the entity for storage.

5.2. Symmetric Cryptographic Module

This section provides the architectural details of symmetric key cryptographic module in the protocol design. The symmetric cryptographic module provides security functionalities that facilitate secure transmission of data between the KGC and the entity, which might be a sensor or a device. The proposed scheme is an attempt to improve the security of ICMetric by generating a symmetric session key between the sensor and the KGC, which is resilient against pre-computed attacks. The generated symmetric key is then used for securely communicating the data between the authenticated parties.

The process of symmetric key generation and authentication is carried out between the entity and the KGC to establish symmetric key for secure data communication between them. The following section details the steps involved in this process:

5.2.1. Strong Symmetric Key Generation

As mentioned above in Section 5.1, the KGC stores certain entries for each registered entity (sensor, device, *etc.*) in the form of a tuple. The entity ID is the identifying attribute for an entity assigned by the KGC at the time of registration. The salt s is generated by the KGC as a random number 128 bit number at the time of registration. The KGC already has a verifier v received and stored for every entity, which it uses now during the process of symmetric key generation and authentication. The entity generated its verifier v at registration time and sent it to KGC as:

$$v = g^{x_A} \, mod \, N \tag{3}$$

where g is a generator of the multiplicative group and n is a safe prime as described in Table 3.

Table 3. Mathematical notations.

Symbol	Meaning
$\|\|$	The concatenation operator
N	A large prime number. All computations are performed modulo n
g	A primitive root modulo n (often called a generator)
$salt_{A,} salt_B$	A random string used as the entity's and KGC's salt
v	The verifier of the entity
IC_A, IC_B	ICMetric of entity and KGC respectively
$ID_{A,} ID_B$	ID of entity and KGC respectively
x_A, x_b	ICMetric number of entity and KGC respectively
a, b	Ephemeral private keys, generated randomly and not publicly revealed
$h()$	One-way hash function e.g., SHA-256
u	Scrambling parameter-A and B concatenated and hashed
K_A, K_S	Session keys

The design of the ICMetric symmetric key generation and authentication scheme is based on the Secure Remote Password Protocol [33,34]. SRP is particularly suitable for this application, since its properties and design works very well with the constraints of ICMetric technology of not transmitting the

ICMetric number. SRP doesn't require the exchange of ICMetric information between parties for the purpose of authentication and key generation.

To formally start the ICMetric based symmetric key generation process each registered sensor that wishes to communicate generates its own ICMetric:

$$x_A = IC_A \tag{4}$$

The authentication process for the purpose of sending data readings to the KGC is initiated by the entity, when it sends an initiation request containing its identification to the KGC asking for the assigned salt value. On contacting the KGC, each sensor receives the salt stored on the KGC under its sensor IDA. Now the sensor generates number a:

$$a = h\,(x_A || \, salt_A \, || \, ID_A) \tag{5}$$

and uses it to calculate A and sends the result to the KGC:

$$A = g^a \; mod \; N \tag{6}$$

The KGC does a similar operation to calculate b:

$$b = h\,(x_B || \, salt_B \, || \, ID_B) \tag{7}$$

thereby calculating B and also adds the public verifier to it and finally sending B to the sensor:

$$B = (kv + g^b) \; mod \; N \tag{8}$$

where $k = h(N||g)$.

Both sides compute a random scrambling parameter u=h (A || B) based on the exchanged A and B. So both sides can now construct the shared session key. The sensor constructs it as follows:

$$\begin{aligned} S_B &= (B - k \cdot v)^{a+ux} \, mod \, N \\ &= (kv + g^b - k \cdot v)^{a+ux} \, mod \, N \\ &= (g^b)^{a+ux} \\ K_A &= h\,(S_A) \end{aligned} \tag{9}$$

The KGC constructs it as follows:

$$\begin{aligned} S_B &= (A \cdot v^u)^b = (g^{a+xu})^b \\ K_B &= h\,(S_B) \end{aligned} \tag{10}$$

Both sides now possess the same secure session key K based on the respective formulae.

5.2.2. Symmetric Key Authentication

To complete the authentication, now the sensor needs to prove to KGC that its key is identical. In order to do so, the sensor constructs the message M1 and sends it to the KGC:

$$M_1 = h\,(\,N \,||\, g \,||\, ID_A \,||salt_A \,||\, A \,||\, B \,||\, K_A\,) \tag{11}$$

The KGC will calculate M2 using its own KS and compare it against the message received from the sensor. If both keys do not match, the authentication fails resulting in refusal of communication request. If the sensor request is authenticated, the data can safely be relayed to the KGC. The summarized protocol is given in Figure 6. After an entity is authenticated and joins the network, its session key is kept in a secure cache and is valid for a set time period.

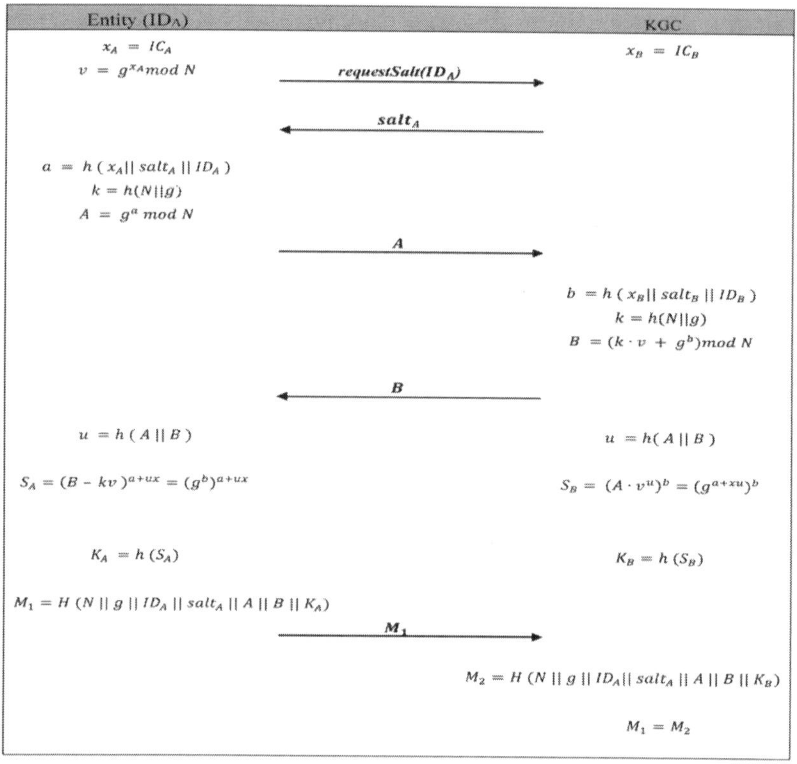

Figure 6. ICMetric based symmetric cryptographic module.

5.2.3. Symmetric Key Encryption/Decryption

Once the session key is in place, secure communications can be carried out using any well-known symmetric key encryption algorithms such as AES or 3DES. Here we propose using AES to securely send health data between the entities and KGC. So in this step of the design we integrate all the individual ICMetric based components with the symmetric encryption/decryption scheme to achieve data confidentiality and integrity between the parties. This module performs encryption/decryption of the health data between the sensors and the KGC.

6. IMPLEMENTATION

A working prototype of the health sensing system was implemented using C on Linux. The scheme has been implemented on a first generation Intel Core i3 3.2 GHz processor with 6 GB RAM. The authentication and key generation module of the design has been implemented using the openssl cryptographic library. For the implementation of symmetric encryption/decryption we have used the CyaSSL library. Although there is support for AES in OpenSSL, but we use the CyaSSL library for implementing the encryption/decryption counterpart of the design. The CyaSSL [35] library is a lightweight SSL library that specifically targets resource constrained devices. CyaSSL is 20 times smaller than the standard OpenSSL. CyaSSL also takes advantage of Intel AES-NI support [36]. This helps boost the performance of AES, since the instructions execute directly on the chip instead of executing the algorithm from software, thereby making it run 5–10 times faster. We intend to deploy this scheme on an embedded system. The design and implementation choices are greatly influenced by the fact that embedded platforms possess limited resources. Detailed results in the upcoming section also show that the scheme can achieve high levels of security without resource demand when deploying on a real sensor test bed.

The authentication and key generation module of the system implementation has five variants namely 160, 224, 256, 384, 512 bits; with a 128-bit device ICMetric as input. To offer flexibility the encryption/decryption module of the scheme has been implemented to facilitate two AES variants namely AES-128 and AES-256.

7. RESULTS AND DISCUSSION

In this section, the efficiency of the design is evaluated by measuring the RAM consumption and execution time of the implemented scheme. For the purpose of evaluating the execution time of the system the programs runtime itself is utilised, whereas for the purpose of evaluating the memory consumed by the prototype Valgrind tool was employed.

Valgrind [37] is a code profiling and memory debugging tool for Linux that employs Massif for memory profiling. Massif [38] analyses the implementation and measures how much memory the program uses during its lifetime and classifies this as follows:

- Useful heap—heap used by the program over time.
- Extra heap—heap blocks allocated for administration data over the program's lifetime.
- Total heap—total of the useful heap and extra heap allocated for the program over its lifetime.

- Stack—stacks used by the program over time.

7.1. Performance Evaluation of ICMetric Key Generation and Authentication Module

7.1.1. Execution Time

The time requirements for authentication and the key generation counterpart, which are responsible for the authentication and the key generation of an ICMetric based symmetric key, are evaluated. A 128 bit ICMetric is utilised as input for testing all the variants of the prototype.

To prove the scalability of the key generation scheme, the time taken for the generation of keys of varying key sizes was evaluated. Figure 7 shows the key size *versus* time taken graph for computing key variants namely 160, 224, 256, 384, 512 bits. It is evident from the graph that an increase in the key size does not bring a drastic change to the time performance of the application. Hence, the scheme is able to provide higher levels of security without substantial execution time overheads.

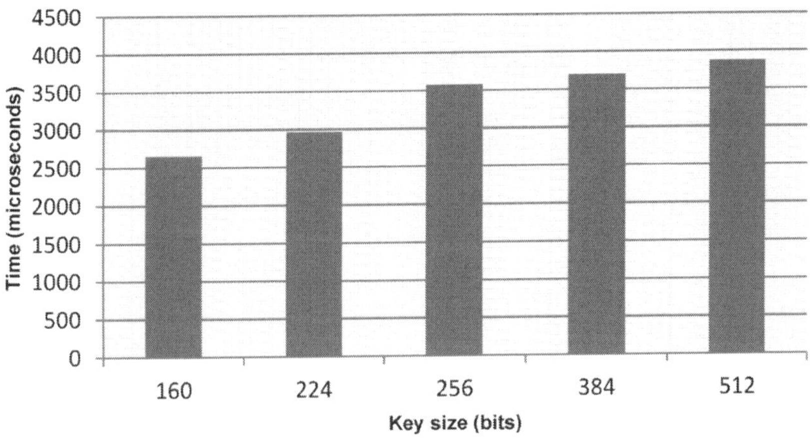

Figure 7. Execution times for ICMetric based symmetric key variants.

7.1.2. RAM Consumption

Here, the memory performance of ICMetric based symmetric key authentication and key generation counterpart has been evaluated. Once again, a 128 bit ICMetric is utilized as input for testing all the variants of the prototype. The memory performance of the ICMetric based authentication and key generation counterpart is evaluated using Valgrind. The generated graphs in Figure 9 depict memory profile of the five ICMetric symmetric key variants, by presenting the program lifecycle and the memory (bytes) consumed.

Figure 9a shows that the maximum total memory consumed for the generation and authentication of shortest 160-bit key is around 10KB; while Figure 8e shows that the maximum total memory consumed for the generation and authentication of longest 512-bit key is around 11 KB. This proves that an increase in the key size has a very small effect on the memory consumption of the application. The graphs in Figure 8 clearly demonstrate that the health sensing scheme presented here, is able to provide high levels of security at the cost of a very limited memory footprint.

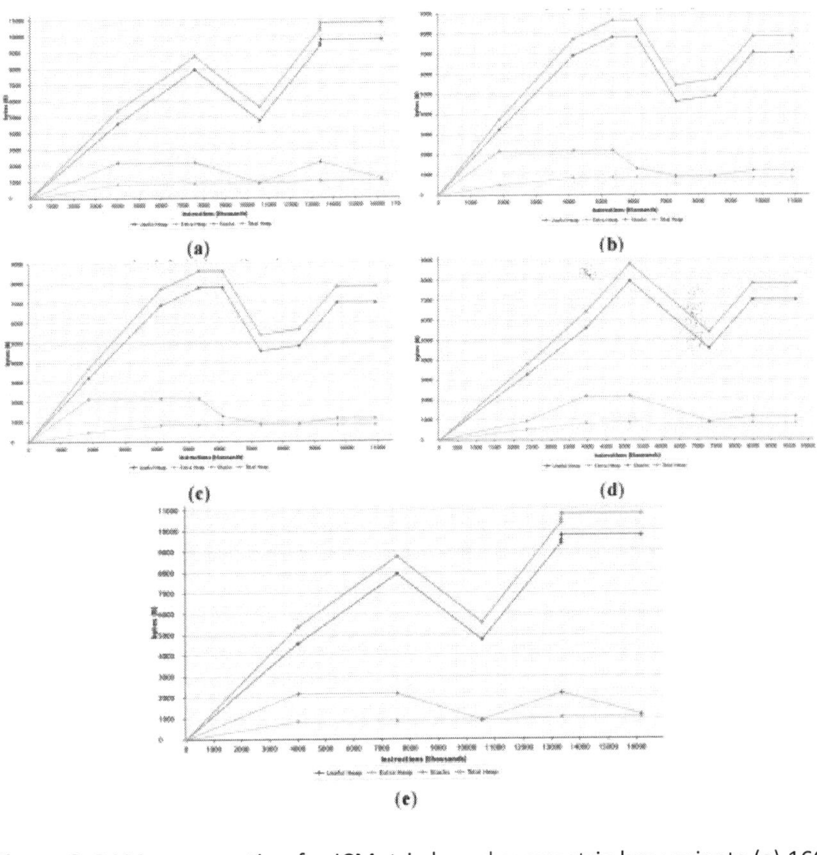

Figure 8. RAM consumption for ICMetric based symmetric key variants (**a**) 160-bit; (**b**) 224-bit; (**c**) 256-bit; (**d**) 384-bit; (**e**) 512-bit.

7.2. Performance Evaluation of ICMetric Based AES Encryption/Decryption Module

7.2.1. Execution Time

In this section, the time performance of encryption/decryption module that is responsible for carrying out the encryption/decryption of a data block using the ICMetric symmetric key is evaluated. Figure 9 shows a bar graph in which the 128 and 256 bit AES key variants *vs.* the time taken. It is evident from the graph that an increase in the key size has a limited effect on the execution time.

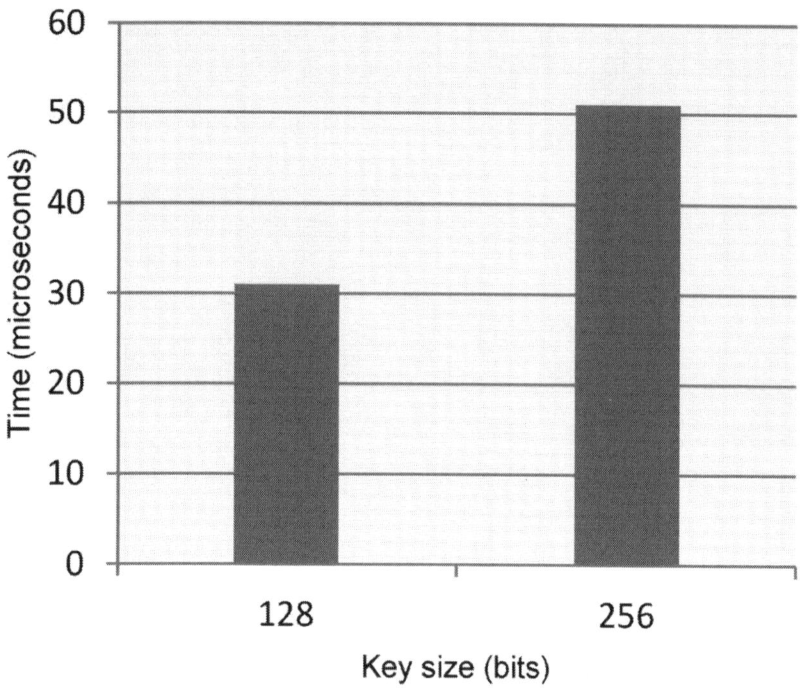

Figure 9. Execution times for ICMetric based AES variants.

7.2.2. RAM Consumption

The graphs in Figure 10a,b present memory profile for the ICMetric based 128-bit and 256-bit AES encryption/decryption, respectively.

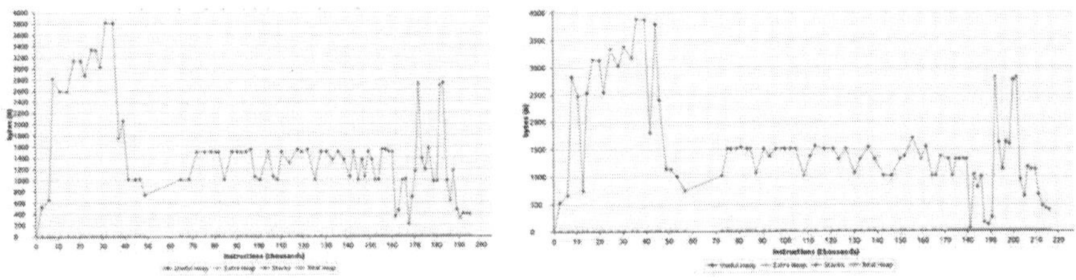

(a) (b)

Figure 10. RAM consumption for ICMetric based AES encryption/decryption (**a**) 128-bit; (**b**) 256-bit.

It is apparent from the graphs in the figure that the maximum total memory consumed for the ICMetric based AES encryption/decryption using 128-bit key is around 3.8 KB and for the 256-bit key is 4 KB, which is very suitable for resource constrained devices. Therefore the proposed scheme is able to provide higher levels of security without substantial memory performance overheads.

7.3. Discussion

In the following section we perform a security analysis of the designed secure health sensing scheme with respect to the security goals accomplished, while also identifying the limitations of the proposed framework.Table 4 gives a summary overview of the goals accomplished/unaccomplished by individual system modules of the health sensing scheme presented in this paper.

Table 4. System modules and goals accomplished.

		System Goals					
		Authentication	Confidentiality	Access Control	Non Repudiation	Precomputed Attacks	Integrity
Modules	MEMS Accelerometer	√					
	ICMetric			√	√		
	Strong Key Generation					√	
	Symmetric Module	√		✓		√	√
	AES		√				√

The health sensing framework combines the security advantages of the device ICMetric and the designed symmetric key generation scheme for the generation of strong symmetric keys between the entities and the KGC. Non repudiation is an inferred security goal of the scheme, since generating the ICMetric from a range of features that are unique to a computation device prevents an entity from denying a particular action performed by their device. The original ICMetric scheme [17] has inherent weaknesses like insufficient length and entropy, owing to which it is not well suited to serve as a key for cryptographic operations. The symmetric key generation scheme in our security framework uses key derivation functions on ICMetric number coupled with a salt, to generate a strong symmetric ICMetric key. Following their use, the ICMetric and the cryptographic keys are discarded hence deterring all forms of key capturing. Each device concatenates its ICMetric with a random per-user value (so called "salt") and stores the hash value of the result along with the salt. This makes certain kinds of brute-force attacks and dictionary attacks more difficult. The security features of the design also prevent the possibility of man in the middle attack. The scrambling parameter in the symmetric cryptographic module ensures that each entity only gets one verification attempt at impersonation, resulting in the ruling out the possibility of man-in the middle attack.

A major advantage provided by the health sensing application is that of authenticating entities without the need for human intervention. This is particularly important from the patient's perspective, since human intervention is not always possible for authentication, therefore the authentication functionality is automatically carried out based on the sensors' ICMetric keys. To generate the symmetric key and carry out the authentication/access control, the framework effectively conveys a zero knowledge password proof. This property is guaranteed by the symmetric cryptographic module and ensures that the entity is able to prove to the KGC about the knowledge of the ICMetric number, without actually transmitting/communicating the ICMetric to the verifying party. The data sent by each entity is encrypted using AES based on the ICMetric symmetric key, so that health data is not leaked to unauthorized parties. Integrity checking is a further functionality provided by the health sensing scheme during the secure transmission of data between the patient and the health professional. As shown in Table 4, the symmetric cryptographic module and the AES module are responsible for ensuring the integrity of computed symmetric key and communicated patient data.

8. CONCLUSIONS

Wearable healthcare devices have revolutionised the way in which healthcare is delivered. Wearable computing has now made it possible to deliver healthcare beyond the confines of the hospital. While there are many advantages of monitoring a person's health remotely, a major hurdle with this healthcare model is the provision of security. In this paper an ICMetric based health sensing scheme is proposed that employs the MEMS accelerometer and other device characteristics to generate a sensor identification called the device ICMetric. This device ICMetric possesses the quality that it is generated without human intervention and only when it is required. After use, the device ICMetric and all related data are removed from the system. In the proposed scheme the functionalities of the ICMetric are combined with a symmetric key protocol for providing access control, authentication and confidentiality of health data in a resource efficient manner. By using the ICMetric technology no form of credentials are residing on a system; thus enabling to prevent major threats like key theft, man-in-the-middle attack and brute force attack related to secure communication of health data. The design decisions made in this framework have also enabled it to be particularly suitable for resource constrained environments, which require elimination of computationally expensive and resource intensive operations while providing the required security. The security analysis and implementation results of this lightweight design prove that the protocol presented in this paper is suited greatly for health monitoring applications in terms of security and resource consumption. In future, we plan to carry out practical evaluation of the proposed security framework on real sensor testbed. The issues related to availability of the proposed scheme are currently not within the scope of this work. Denial of service attacks target the availability of a system, thus future plans include also addressing this goal.

ACKNOWLEDGMENTS

This research has been supported by the Engineering and Physical Sciences Research Council (EPSRC), Grant EP/K004638/1 (project named RoBoSAS).

AUTHOR CONTRIBUTIONS

Ruhma Tahir was responsible for conceiving the scheme. She also tailored the scheme so that it works with the ICMetric technology. The ICMetric generation scheme was designed and simulated by Hasan Tahir. He also performed a security analysis of the scheme and its simulation results. Klaus McDonald-Maier provided major guidance and input in the design of the scheme so that it can be customized according to the requirements of the health domain. The manuscript was prepared, studied and deliberated upon by all authors.

REFERENCES

1. Vandrico Inc. (2008) Wearable Technology Database. Available online: http://vandrico.com/database(accessed on 13 October 2015).
2. Zheng, Y.L.; Ding, X.R.; Poon, C.C.Y.; Lo, B.P.L.; Zhang, H.; Zhou, X.-L.; Yang, G.Z.; Zhao, N.; Zhang, Y.-T. Unobtrusive sensing and wearable devices for health informatics. *IEEE Trans. Biomed. Eng.* **2014**, *26*, 1538–1554.
3. Kovalchuk, Y.; Howells, G.; McDonald-Maier, K. Overview of ICMetrics technology—Security infrastructure for autonomous and intelligent healthcare system. *Intern. J. Serv. Sci. and Technol.* **2014**, *4*, 49–60.
4. Tahir, R.; Hu, H.; Gu, D.; McDonald-Maier, K.; Howells, G. A Scheme for the Generation of Strong ICMetric Based Session Key Pairs For Secure Embedded System Applications. In Proceeding of the International Conference on Advanced Information Networking and Applications, Barcelona, Spain, 25–28 March 2013; pp. 689–696.
5. Alam, M.M.; Hamida, E.B. Surveying wearable human assistive technology for life and safety applications: standards, challenges and opportunities. *Sensors* **2014**, *14*, 9153–9209.
6. Bourbakis, N.; Pantelopoulos, A.; Kannavara, R.; Nikita, K. *Security and Privacy in Biomedical Telemetry: Mobile Health Platform for Secure Information Exchange*; Handbook of Biomedical Telemetry Wiley IEEE: Hoboken, NJ, USA, 2014; pp. 382–418.
7. Ragesh, G.K.; Baskaran, K. CRYP: Towards Cryptographically Enforced and Privacy Enhanced WBANs. In Proceeding of the First International Conference on Security of Internet of Things, Kollam, India, 17–19 August 2012; pp. 204–209.
8. Yi, X.; Willemson, J.; Nait-Abdesselam, F. Privacy-Preserving Wireless Medical Sensor Network. In Proceeding of the 12th IEEE International Conference on Trust, Security and Privacy in Computing and Communications, Melbourne, Australia, 16–18 July 2013; pp. 118–125.

9. Barua, M.; Liang, X.; Lu, R.; Shen, X. PEACE: An Efficient and Secure Patient-Centric Access Control Scheme for eHealth Care System. In Proceedings of the IEEE Conference on Computer Communications Workshops, Shanghai, China, 10–15 April 2011; pp. 970–975.
10. Rostami, M.; Juels, A.; Koushanfar, F. Heart-to-Heart (H2H): Authentication for Implanted Medical Devices. In Proceeding of the 20th ACM Conference on Computer and Communications Security, Berlin, Germany, 4–8 November 2013; pp. 1099–1112.
11. Kannavara, R.; Mertoguno, S.; Bourbakis, N. SCAN secure processor and its biometric capabilities. *J. Electron. Imaging* **2011**, *20*.
12. Venkatasubramanian, K.K.; Banerjee, A.; Gupta, S.K.S. PSKA: Usable and secure key agreement scheme for body area networks. *IEEE Trans. Inf. Technol. Biomed.* **2010**, *1*, 60–68.
13. Jurik, D.; Weaver, A.C. Securing Mobile Devices with Biotelemetry. In Proceeding of the 20th International Conference on Computer Communications and Networks, Maui, Hawai, 31 July–4 August 2011; pp. 1–6.
14. Venkatasubramanian, K.K.; Banerjee, A.; Gupta, S.K.S. EKG—Based Key Agreement in Body Sensor Networks. In Proceeding of the 2nd Workshop on Mission Critical Networks, Phoenix, AZ, USA, 13–18 April 2008; pp. 1–6.
15. Strydis, C.; Seepers, R.M.; Peris-Lopez, P.; Siskos, D.; Sourdis, I. A System Architecture, Processor and Communication Protocol for Secure Implants. *ACM Trans.* **2013**, *10*, 1–23.
16. Hu, C.; Xiuzhen, C.; Zhang, F.; Wu, D.; Liao, X.; Chen, D. OPFKA: Secure and Efficient Ordered Physiological Feature Based Key Agreement for Wireless Body Area Networks. In Proceedings of the 32nd IEEE International Conference on Computer Communications, Turin, Italy, 14–19 April 2013; pp. 2274–2282.
17. Papoutsis, E. Investigation of The Potential of Generating Encryption Keys For ICMetrics. Ph.D. Thesis, The University of Kent, Canterbury, UK, June 2009.
18. Kokosy, A.; Floquet, T.; Howells, G.; Hu, H.; Pepper, M.; Sakel, M.; Donze, C. SYSIASS—An Intelligent Powered Wheelchair. In Proceedings of the 1st International Conference on Systems and Computer Science, Lille, France, 29–31 August 2012.
19. Defence Advanced Research Project Agency; Micro Electro-Mechanical Systems (MEMS). *Annu. Rev. Fluid Mechan.* **1998**, *30*, 579–612.

20. Castoldi, L. The MEMS Revolution (2012, September). Available online: http://www.semi.org/eu/sites/semi.org/files/docs/STM.pdf (as sessed on 15 October 2015).
21. Kuc, R. *Electrical Engineering in Context: Smart Devices, Robots & Communications*; Cengage Learning: Boston, MA, USA, 2014.
22. Feng, M.; Fukuda, Y.; Mizuta, M.; Ozer, E. Citizen sensors for SHM: Use of accelerometer data from smartphones. *Sensors* **2015**, *15*, 2980–2998.
23. Dey, S.; Roy, N.; Xu, W.; Choudhury, R.R.; Nelakuditi, S. AccelPrint: Imperfections of Accelerometers Make Smartphones Trackable. In Proceedings of the 21st Annual Network and Distributed System Security Symposium NDSS, San Diego, CA, USA, 23–26 February 2014.
24. Mobile Device Identification via Sensor Fingerprinting. National Research & Simulation Center, Rafael, Internal Report 2014. Available online: https://crypto.stanford.edu/gyrophone/sensor_id.pdf (assesse d on 15 October 2015).
25. Aysu, A.; Ghalaty, N.F.; Franklin, Z.; Yali, M.P.; Schaumont, P. Digital Fingerprints for Low-Cost Platforms Using MEMS Sensors. In Proceedings of the Workshop on Embedded Systems Security, Montreal, Canada, 29 September 2013.
26. Burns, A.; Greene, B.R.; McGrath, M.J.; O'Shea, T.J.; Kuris, B.; Ayer, S.M.; Stroiescu, F.; Cionca, V. SHIMMER™—A wireless sensor platform for noninvasive biomedical research. *IEEE Sens. J.* **2010**, *9*, 1527–1534.
27. Freescale Semiconductor. (2008, April) ±1.5g, ±6g Three Axis Low-g Micromachined Accelerometer. Technical Data. Available online: http://www.freescale.com/files/sensors/doc/data_sheet/MM A7361L.pdf(assessed on 10 October 2015).
28. Maxim Integrated. DS2411 Silicon Serial Number with VCC Input (2003, May). Data Sheet. Available online: http://datasheets.maximintegrated.com/en/ds/DS2411.pdf (as sessed on 10 October 2015).
29. Ferraris, F.; Grimaldi, U.; Parvis, M. Procedure for effortless in-field calibration of three-axial rate gyro and accelerometers. *Sens. Mater.* **1995**, *7*, 311–330.
30. Tahir, H.; Tahir, R.; McDonald-Maier, K. Securing MEMS Based Sensor Nodes in the Internet of Things. In Proceedings of the Sixth International Conference on Emerging Security Technologies, Braunschweig, Germany, 3–15 September 2015.

31. Almulhem, A. Threat modeling for electronic health record systems. *J. Med. Syst.* **2012**, *36*, 2921–2926.

32. Wu, T. The Secure Remote Password Protocol. In Proceedings of the 1998 Internet Society Network and Distributed System Security Symposium, San Diego, CA, USA, 11–13 March 1998; pp. 97–111.

33. RFC 5054, Using the Secure Remote Password (SRP) Protocol for TLS Authentication. November 2007. Available online: http://www.rfc-editor.org/info/rfc5054 (accessed on 10 October 2015).

34. CyaSSL User Manual. User Manual (2013, September). Available online: http://www.yassl.com/documentation/CyaSSL-Manual.pdf (accessed on 10 October 2015).

35. Gueron, S. Intel Advanced Encryption Standard (AES) New Instructions Set. *Intel. Corp. White Pap.* **2012**, *3*, 1–95.

36. The Valgrind Quick Start Guide. User Manual (2014, September). Available online: http://valgrind.org/docs/manual/valgrind_manual.pdf (accessed on 10 October 2015).

37. Wolff, M. (2011, August) Massif-Visualizer Memory Profiling UI. Presentation. Available online: https://desktopsummit.org/sites/www.desktopsummit.org/files/massif-visualizer.pdf (assessed on 15 October 2015).

38. Electronic Authentication Guideline—Recommendations of the National Institute of Standards and Technology (2006, April). Available online: http://csrc.nist.gov/publications/nistpubs/800-63/SP800–63V1_0_2.pdf (accessed on 10 October 2015).

Chapter 13

Integrated Microwave Photonic Isolators: Theory, Experimental Realization and Application in a Unidirectional Ring Mode-Locked Laser Diode

Martijn J.R. Heck [1,*], Sudharsanan Srinivasan [2], Michael L. Davenport [2] and John E. Bowers [2]

[1]Department of Engineering, Aarhus University, Aarhus 8200, Denmark
[2]Department of Electrical and Computer Engineering, University of California Santa Barbara, Santa Barbara 93106-9560, CA, USA

ABSTRACT

A novel integrated microwave photonic isolator is presented. It is based on the timed drive of a pair of optical modulators, which transmit a pulsed or oscillating optical signal with low loss, when driven in phase. A signal in the reverse propagation direction will find the modulators out of phase and, hence, will experience high loss. Optical and microwave isolation ratios were simulated to be in the range up to 10 dB and 20 dB, respectively, using parameters representative for the indium phosphide platform. The experimental realization of this device in the hybrid silicon platform showed microwave isolation in the 9 dB–22 dB range. Furthermore, we present a design study on the use of these isolators inside a ring mode-locked laser cavity. Simulations show that unidirectional operation can be achieved, with a

30–50-dB suppression of the counter propagating mode, at limited driving voltages. The potentially low noise and feedback-insensitive operation of such a laser makes it a very promising candidate for use as on-chip microwave or comb generators.

Keywords: hybrid silicon platform; microwave photonics; mode-locked laser diodes; optical isolators; photonic integrated circuits; semiconductor optical amplifiers; silicon photonics

1. INTRODUCTION

The field of microwave photonics (MWP) utilizes photonics to generate, process, transport and detect radio frequencies (RF), typically in the gigahertz range [1]. This approach benefits from the large bandwidth and low loss signal propagation of photonic components, as compared to electronic components and circuits. Examples of successful MWP applications include high-bandwidth wireless links operating at 100 Gbps [2], low noise microwave oscillators [3], clock-free data generation [4] and photonic analog-to-digital converters for high bandwidth operation [5].

Such MWP systems are typically built using fiber-optic and discrete components, limiting their real-world applications, due to lack of robustness, high cost and large size and weight. Integration of MWP systems on a photonic integrated circuit (PIC) can in principle address these concerns and limitations. PIC technology has now evolved to the level that industry and academia have access to mature technology platforms [6]. This led to the field of integrated MWP, where MWP functions are integrated on a PIC [7]. Recent advances, like the ultra-low waveguide losses, realized in the silicon nitride platform, can further unlock this potential, allowing for the almost fiber-like performance of a PIC-based system, e.g., for low noise microwave generation [8].

A main bottleneck for the integration of larger-scale systems on a PIC is the lack of practical integrated optical isolators, e.g., to prevent optical feedback from entering an on-chip laser cavity. For high-end MWP applications, low noise performance is of key importance, but such feedback results in increased phase and relative-intensity noise, increased optical linewidths or even chaotic laser emission [9,10]. Research efforts to realize an integrated isolator are mainly based on the integration of magneto-optic materials onto the PIC platform, to enable a non-reciprocal absorption or phase shift, however with a limited extinction ratio and high insertion loss [11]. Another way to break the symmetry between the forward and backward signals is by applying a time-dependent electro-optic index or transmission modulation, as presented in, e.g., [12,13]. More recently, isolation based on acousto-optic modulation was also shown, though without a clear path towards integration on mature PIC platforms [14].

The time-based asymmetry of a pulsed or MWP signal, propagating through a sequence of optical gates, which open and close periodically, can be used to achieve a difference in loss between forward and backward propagating signals. If a forward propagating oscillating signal passes through an open gate when its amplitude is maximum, its loss is minimal. However, if the spacing and timing between the gates is designed well, a backward propagating signal will not find all gates open and, hence, will incur additional losses. This general concept was theoretically studied in [15], where the self-amplitude modulation of a set of saturable absorption modulators (AMs) was used to create isolation between a pulsed signal and lower power feedback or noise.

In this work, we use a conceptually similar approach, but drive the modulators actively, by applying a time-varying voltage, effectively creating periodic gates. The concept and the simulation of the MWP isolator using realistic parameters, representative for indium phosphide-based active components, will be presented in Section 2. The realization in the hybrid silicon platform will be described in Section 3. Experimentally obtained results will be shown and discussed. In Section 4, we will theoretically investigate a possible promising implementation of the MWP in a ring mode-locked laser (MLL). It will be shown that unidirectional operation of the ring MLL can be achieved. To conclude, in Section 5, we will discuss the wider implementation and application range of these concepts.

2. MWP ISOLATOR CONCEPT AND THEORETICAL STUDY

The concept for an MWP isolator is shown in Figure 1, using two modulators. The modulators are driven with a period *T*, by varying the voltage over them, and, hence, have a time-dependent transmission. A forward propagating oscillating signal in phase with the transmission modulation of the modulators will experience minimum loss. This figure shows clearly how a backward propagating signal will always encounter at least one modulator in a (partially) blocking state, if the distance between the two modulators corresponds to $0.25 \cdot T \cdot v_g$, with v_g the group velocity of the signal in the system. This means that from the perspective of the backward propagating signal, the two modulator gates are always modulated out of phase with each other, whereas from the perspective of the forward propagating signal, these modulator gates are modulated in phase.

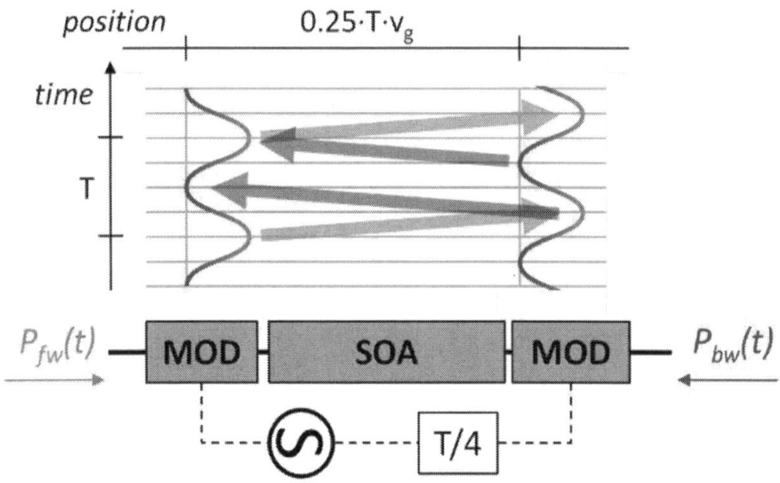

Figure 1. Schematic of the microwave photonics (MWP) isolator concept using modulators (MODs), showing the concatenated MOD-semiconductor optical amplifier (SOA)-MOD array, with the forward propagating signal P_{fw}entering from the left and the backward propagating signal P_{bw} entering from the right. The MOD gates are driven with a *T*/4 time difference, with *T*the period of the RF signal. The MOD transmission is indicated schematically with the sinusoid, with period *T*, with red indicating low transmission and green indicating high transmission. The green and red arrows are following the time-position grid and show how a forward propagating signal (green arrow) can pass two open gates, whereas a backward propagating signal (red arrow) always encounters at least on (partially) closed gate. The optical amplifier (SOA) can be used to compensate for overall insertion losses.

Although this concept would work with any type of RF-driven modulator, we limit the discussion to reversely-biased semiconductor optical amplifier (SOA) sections, effectively creating AMs, as was done in [12]. The main reason is the obvious compatibility with the amplifying SOA sections in a PIC, which allows for integration of the MWP isolator after the output of a laser and even into a laser cavity, as will be shown in Section 4. As mentioned above, this is the main rationale for an integrated isolator. From a fabrication perspective, this means that no epitaxial regrowth or overgrowth steps are required, nor techniques like quantum-well intermixing. Experimentally, it was validated that the SOA transmission, more specifically the absorption, changes with applied reverse bias voltage, and hence, it is feasible to use it as an AM [16]. Note that in this implementation, the operating wavelength is well below the bandgap wavelength of the AM, which is different from what is typically called an electro-absorption modulator (EAM), in which case the wavelength is close to the bandgap.

We can model the MWP isolator, as shown in Figure 1, using SOA rate equations for the optical field $A(z,t)$ and the carrier density $N(z,t)$ inside the SOA as presented in, e.g., [12,17]:

$$\frac{\partial A(z,t)}{\partial z} \pm \frac{1}{v_g}\frac{\partial A(z,t)}{\partial t} = \frac{1}{2}\frac{G(z,t)}{1+\varepsilon P(z,t)}A(z,t) - \frac{i}{2}\alpha_N G(z,t)A(z,t) - \frac{1}{2}\alpha_{int}A(z,t) + F_{SE}(z,t) \quad (1)$$

$$\frac{\partial N(z,t)}{\partial t} = \frac{I}{qV} - \frac{N(z,t)}{\tau} - \frac{1}{\Gamma\hbar\omega\sigma}\frac{G(z,t)P(z,t)}{1+\varepsilon P(z,t)} \quad (2)$$

with the gain $G(z,t)$ expressed as a linearized dependency on the carrier density $N(z,t)$:

$$G(z,t) = \Gamma a_N\left(N(z,t) - N_{tr}\right) \quad (3)$$

Here, the optical field $A(z,t)$ is expressed in [$W^{1/2}$], $P(z,t)$ is the optical power in [W], ε is the nonlinear gain compression (0.2 W^{-1} for SOA, 0 for AM, *i.e.*, reverse bias), α_N is the linewidth enhancement factor (3 for SOA, 1 for AM), α_{int} = 20 cm^{-1} the passive loss, $F_{SE}(z,t)$ a Langevin noise term representing spontaneous emission, I the injection current, q the elementary charge, V the volume of the active area, τ the carrier recovery time (300 ps for SOA, 5 ps for AM), Γ = 0.3 the confinement factor, $\hbar$ is Planck's constant, ω the optical angular frequency, σ the mode cross section, a_N the differential gain coefficient (3×10^{-20} m^2 for SOA, 12×10^{-20} m^2 for AM) and N_{tr} the transparency carrier density (0.3×10^{18} cm^{-3} for SOA, 0.7×10^{18} cm^{-3} for AM). These parameters are representative for the indium phosphide technology and have been experimentally verified, e.g., [13]. The absorption dependency of the AM on the bias voltage is modeled using a voltage-dependent transparency carrier density $N_{tr} = N_{tr}(V(t))$, as proposed in [18] and experimentally verified in [13]. These equations are numerically solved using a spatial resolution of $\Delta z = 5$ µm, corresponding to a time resolution of $\Delta t = 60$ fs.

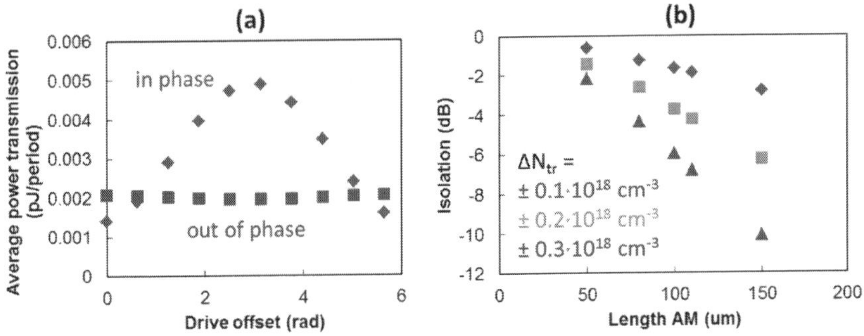

Figure 2. (**a**) Power transmission of a 1-mW peak power 20-GHz MWP signal through an isolator with a total length of 1.2 mm and with two absorption modulator (AM) sections of 100 µm. The bias amplitude corresponds to $\pm 0.2 \times 10^{18}$ cm^{-3}. Referring to Figure 1, the forward transmission (red) and reverse, backward transmission (blue) are shown, as a function of the phase difference between the MWP signal and the AM driving voltage. (**b**) Maximum difference between forward and backward propagating signals for various AM lengths and driving voltages (indicated with N_{tr}).

Using this model, the feasibility of the MWP isolator can be explored. The transmission of the forward and backward propagating 20-GHz MWP fields for an isolator with 100 µm-long AMs is shown in Figure 2a. Obviously, the AM driving voltage needs to be in phase with the MWP signal oscillator for maximum transmission, as shown by the maximum around three radians drive offset. Note that this is a matter of calibrating the model, or any real-world system, and represents a relative value only. Since a system integrated on a PIC has known path lengths, the required drive offset is known beforehand. On the other hand, the counter-propagating signal has in practice an unknown phase with respect to the AMs, as reflections can originate from anywhere downstream of the circuit. However, as can be seen in Figure 2a, this transmission is flat and below the peak transmission of the forward signal, indicating the isolation capability. Note that this flat transmission also means that isolation from continuous-wave counter-propagating signals is achieved.

In Figure 2b, the isolation is plotted for various AM lengths and various driving voltage amplitudes of 3–9 V (corresponding to N_{tr} = (0.1–0.3) × 10^{18} cm^{-3}; [13]). As can be expected, isolation increases with AM length and driving voltage amplitude. However, it has to be noted that such driving voltages are quite high, requiring between 100 mW and 1 W of RF power on the AM, assuming a 50-Ω termination. Based on these results, it can be seen that modest optical isolation ratios up to 10 dB can be achieved, which corresponds to close to a 20-dB microwave isolation, as most of the power is in the 20-GHz microwave fundamental harmonic, with up to 20% in the 40-GHz harmonic, due to the nonlinear transmission.

3. EXPERIMENTAL REALIZATION IN HYBRID SILICON TECHNOLOGY

The MWP isolator was fabricated in the hybrid silicon platform [19]. In this platform, an indium phosphide layer is bonded to a processed silicon photonic chip and then processed further, ensuring lithographic alignment accuracy. Silicon photonic components, such as waveguides, splitters and multiplexers, can be combined in this platform with hybrid waveguides, where the optical mode overlaps with the indium phosphide layers. In this way, active optical functionalities, such as gain, can be achieved. A schematic of the device cross-section is shown in Figure 3a, where it can be seen that the optical mode propagates through an AM-SOA-AM sequence. The tapered mode converters from the silicon waveguide to the hybrid waveguide are electrically pumped, hence the small SOA sections at the start and end of the device [20]. For details about the fabrication process, we refer to [16]. The chip is diced, polished and anti-reflection coated for characterization.

The experimental setup to characterize the MWP isolator is shown in Figure 3c. An MWP signal is generated by an oscillator that is synchronized with the AM drivers. Two microwave phase shifters allow for independent tuning of the phase of the driving voltage on the AMs, compared to the MWP signal. The optical carrier wavelength is 1583 nm, as the AM has maximum dynamic range here, expressed as absorption change as a function of applied voltage. This was determined by looking at the amplified spontaneous emission spectrum of the SOA through the AM, on which the bias was varied, as shown in Figure 4a. The MWP isolator insertion loss is 4.4 ± 0.5 dB, which of course depends on the SOA bias, in this case 200 mA. The overall active device length is 1.4 mm, with two 100-μm long AMs, positioned with a 1.0-mm spacing.

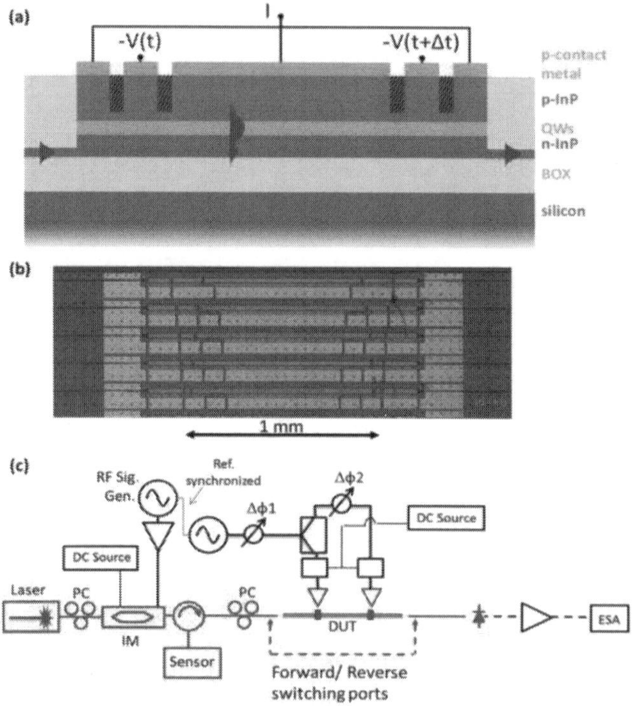

Figure 3. (**a**) Schematic of the hybrid silicon MWP isolator, showing the optical field (red) as it propagates through the silicon waveguide, hybrid silicon III/V waveguide and the silicon waveguide again. Hydrogen-implantation of the p-doped InP cladding provides electrical isolation between the forward-biased SOA and the reverse-biased AM. N-contacts are to the side and not shown in this cross-section. (**b**) Microscope picture of an array of realized devices. (**c**) Experimental setup, showing the MWP signal generation using an intensity modulator (IM, V_π = 3.6 V), driven by a signal generator (RF Sig. Gen.). This signal generator is synchronized with another signal generator that drives the two AMs. Two microwave phase shifters ($\Delta\varphi_{1,2}$) are used to control the phase. The signal is recorded using an electrical spectrum analyzer (ESA) connected to a 40-GHz bandwidth photodiode and a 20-dB RF amplifier. Polarization controllers (PCs) are used to ensure transverse-electric polarization of the field inside the device under test (DUT). Lensed fibers are used to couple the light in and out of the device. Electrical connections are shown in black, optical connections in red. An estimated power of 12.5 dBm is used to drive each of the AMs. The SOA is biased at 200 mA. The circulator with the sensor is used for fiber to chip alignment.

Figure 4b shows the 20-GHz MWP signal transmission with $\Delta\varphi_1$ set to drive one AM in phase with the MWP signal, while tuning the phase of the voltage on the other AM by $\Delta\varphi_2$ (Figure 3b). As can be seen, the received RF power varies with changing the phase on the second AM. Using this plot as a calibration, it can be seen that at $\Delta\varphi_2 = 50° \pm 10°$, the two AMs are driven at $\Delta T = T/4$, as explained in Figure 1, with $T = 50$ ps. This means maximum transmission for the forward signal and minimum transmission for the counter-propagating signal. At $\Delta\varphi_2 = 230° \pm 10°$, this situation is reversed, as expected, and represents $\Delta T = -T/4$. From this plot, it can be seen that 22-dB and 9-dB microwave isolations have been achieved, which agrees well with the optical isolation ratios up to 10 dB, obtained using our simulation. This comparison should be considered qualitative, as a complete set of SOA and AM parameters for the hybrid silicon platform is not available, and the simulation used parameters from the monolithic indium phosphide technology. However, since the active part of the hybrid waveguide is indium phosphide and since the simulation tool uses parameter values experimentally obtained with an indium phosphide SOA, it can be assumed that these outcomes should be close.

We note that ideally, the operation of the device should be symmetric for forward and reverse propagation. From Figure 4b, it can be seen that this is not the case in the experiments. We hypothesize that due to processing non-uniformities, most notably variations in contact resistance over the length of the device, the AMs do not operate identically. In the linear regime, that would still not matter, but as the AM and SOA are nonlinear devices, *i.e.*, their transmission is power dependent, a difference between the two AMs would lead to a difference between the forward and reverse signals, even when both AMs are driven in phase.

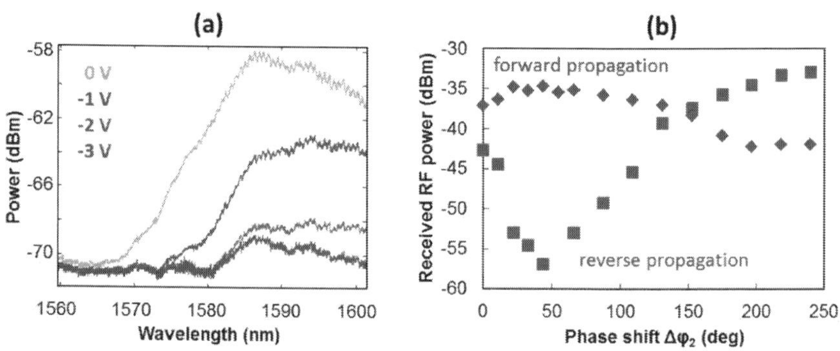

Figure 4. (**a**) Amplified spontaneous emission spectra, obtained by biasing the SOA with a constant current and measured after propagation through an AM.

Bias voltage on the AM is varied from 0–3 V. (**b**) The RF power measured with the electrical spectrum analyzer for the forward and backward propagating MWP signals as a function of the phase difference between the two AM drive voltages.

4. APPLICATION OF THE MWP ISOLATOR IN UNIDIRECTIONAL RING MODE-LOCKED LASERS

The theoretical and experimental results show clearly the potential for isolation, though with a somewhat limited extinction ratio up to ~10 dB optical and ~20 dB electrical. One possible application could be to use this MWP isolator inside a ring laser cavity. It is well known that a small periodic variation in the gain can cause mode-locking, *i.e.*, the laser operates in a pulsed mode [21]. Such MLLs have been realized in various configurations, but ring lasers are particularly attractive due to the lack of mirrors and, hence, the ease of integration on a PIC. However, such ring MLLs are intrinsically bi-directional [18], which leads to lower output efficiency and higher amplitude and phase noise due to cross-gain modulation. It has been shown theoretically and experimentally that by careful MLL cavity design, the power ratio in two counter propagating fields can be controlled, however only to a limited extent [22,23].

In this section, we explore theoretically the feasibility of using the MWP isolator to force the MLL into a unidirectional regime. We use the model presented above (Equations (1)–(3)), extended with a 5-THz SOA gain bandwidth filter, to limit the pulse shortening. The model is implemented using a similar approach as in [19].Figure 5a shows the layout of a ring MLL with a dual AM-phased drive, which means that the MLL is effectively hybridly mode-locked, as the AMs are saturable absorbers and this saturation contributes to the pulse shaping mechanism. The length of the ring determines the period T of the output pulses, and the AMs are placed with a separation corresponding to $T/4$. Figure 5b,c shows the simulated clockwise (CW) and counter-clockwise (CCW) optical output fields for a 20-GHz ring MLL and with an AM drive delay of 12.5 ps and −12.5 ps, respectively. As can be seen, the ring MLL operates unidirectionally with the correct phase of the drive voltage on the AMs. The isolation or suppression of the counter propagating field depends on the MLL operating conditions. In Figure 6, this is shown for a MLL with varying drive voltage amplitude, here again expressed by varying the amplitude of the oscillation of the transparency carrier density N_{tr}. The suppression of the counter propagating field ranges from 30 dB–50 dB, for increasing drive voltage amplitude.

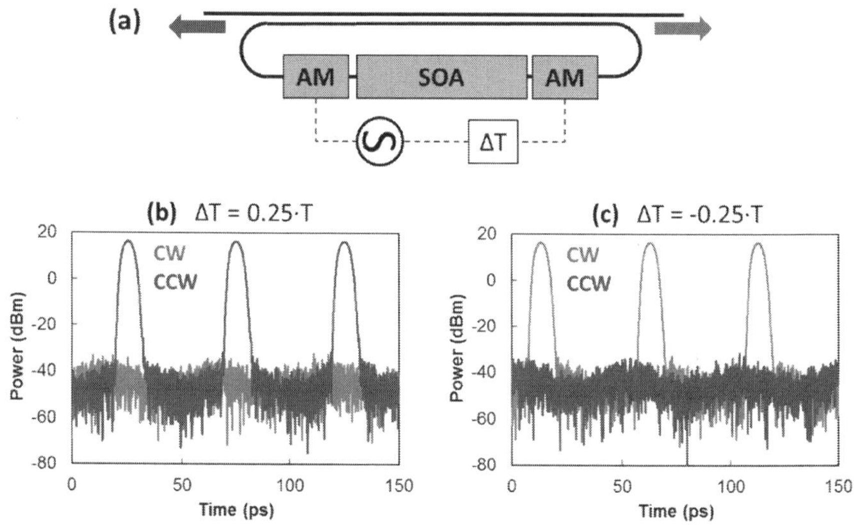

Figure 5. (**a**) Schematic layout of the unidirectional mode-locked laser (MLL), showing an oscillator driving the two AMs, with a relative delay of ΔT. A directional coupler couples light out of the ring into the clockwise (CW, blue) or counterclockwise (CCW, red) output waveguide. The simulated output of a 20-GHz MLL is shown for (**b**) ΔT = 12.5 ps and (**c**) ΔT = −12.5 ps. The SOA length is 800 µm, and the AMs have a length of 25 µm. Power output coupling is 0.5 per roundtrip. The simulation parameters are $\Delta N_{tr} = 0.7 \times 10^{18}$ cm^{-3} and I_{SOA} = 80 mA.

In conclusion, we have shown the feasibility of an integrated unidirectional ring MLL, using our experimentally-verified concept of an MWP isolator. Apart from a more efficient output, the lack of cross-gain modulation and spatial hole burning will likely lead to lower phase and amplitude noise. Furthermore, it can be hypothesized that the laser is less sensitive to optical feedback, as this feedback is injected in the suppressed counter propagating mode. Major applications can be foreseen in on-chip low noise microwave generation and in on-chip comb generation for wavelength-division multiplexed systems.

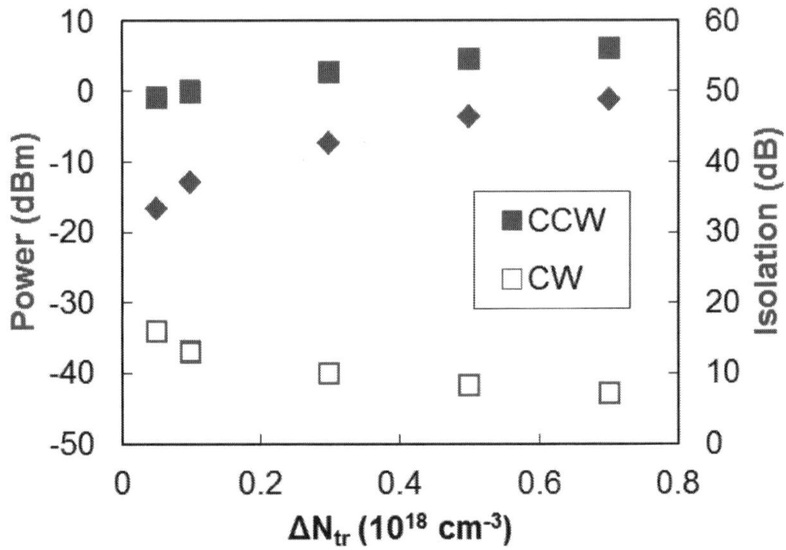

Figure 6. Average output power for the clockwise (CW) and counter-clockwise (CCW) signals, assuming ΔT = 12.5 ps in Figure 5a, as a function of the modulation amplitude ΔN_{tr} of N_{tr}. The isolation ratio is also shown (blue). The SOA injection current is 80 mA. Further MLL design parameters are as in Figure 5.

5. CONCLUSION AND OUTLOOK

In this paper, we have presented the concept of a novel integrated MWP isolator and its first experimental realization in the hybrid silicon platform. Optical and microwave isolation ratios were simulated to be in the range up to 10 dB and 20 dB, respectively, using parameters representative for the indium phosphide platform. The experimental realization of this device in the hybrid silicon platform showed microwave isolation in the 9 dB–22 dB range.

Such isolation ratios are barely sufficient for on-chip use, for example to shield a laser from optical feedback. As was argued in [24], optical isolation of at least 20 dB is required. Options to enhance the MWP isolator include the concatenation of more AM gates and using more square-shaped driving pulses, which would also reduce higher harmonic generation.

In an alternative implementation, when using this MWP isolator inside a resonant cavity, the isolating effect is strongly enhanced. We have presented a design study on the use of MWP isolators inside a ring MLL cavity. Simulations show that unidirectional operation can be achieved, with a 30–50-dB suppression of the counter propagating mode, at limited driving voltages. The potentially low noise and feedback-insensitive operation of such a laser makes it a very promising candidate for use as on-chip microwave or comb generators.

The simulations were based on the indium phosphide material system, and an implementation using reversely-biased SOA sections or AMs was chosen. This choice offers a practical path to integration with (forward biased) SOAs, for example to realize the unidirectional ring MLLs that were presented in this work. Alternative approaches for the AMs can include Mach–Zehnder interferometer modulators in the silicon photonics platform, electro-absorption modulators in the indium phosphide and hybrid silicon platforms or even forward-biased SOAs as gates. The choice will depend on all trade-offs that are required for PIC design and fabrication. Choosing modulators with a higher extinction ratio will improve the isolation ratio. Another important aspect is the optical bandwidth that is required. Mach–Zehnder interferometer modulators and forward-based SOAs are broadband, having bandwidths of at least tens of nanometers. Electro-absorption modulators and reversely-biased SOAs have a very wavelength-dependent transmission and are hence narrow band, typically only a few nanometer, e.g., as shown in [25] and in Figure 4.

The experimental realization on the hybrid silicon platform is a relevant step, as this platform has been shown to have interesting applications in the field of MWP [26]. Although one can argue that similar PIC functionalities can be achieved using a monolithic indium phosphide platform, the hybrid silicon platform offers the potential of economies of scale, in terms of fabrication on 6" or even 8" substrate wafers and the compatibility with the mature CMOS manufacturing infrastructure. This also includes the required integration with electronic driving and control circuitry, which can be done in a package [27] or fully monolithically on the same chip [28]. Such an approach is necessary for large volume applications that can be envisioned for MWP, which include, for example, 5G wireless communications infrastructure.

ACKNOWLEDGMENTS

The authors would like to thank Greg Fish and Erik Norberg from Aurrion Inc. for helpful discussions. This work was supported by the DARPA EPHI Project. M.J.R.H. acknowledges partial funding from Det Frie Forskningsråd through the mmW-SPRAWL project (DFF-4005-00246).

AUTHOR CONTRIBUTIONS

M.J.R.H. conceived of the idea and did the simulations and mask design. M.L.D. developed the fabrication process and fabricated the devices. S.S. performed the experimental characterization. M.J.R.H. and J.E.B. supervised the project. M.J.R.H. wrote the paper, with input from all other co-authors.

REFERENCES

1. Yao, J. Microwave photonics. *J. Lightw. Technol.* **2009**, *3*, 314–335.
2. Pang, X.; Caballero, A.; Dogadaev, A.; Arlunno, V.; Borkowski, R.; Pedersen, J.S.; Deng, L.; Karinou, F.; Roubeau, F.; Zibar, D.; *et al.* 100 Gbit/s hybrid optical fiber-wireless link in the W-band (75–110 GHz). *Opt. Express* **2011**, *19*, 24944–24949.
3. Yao, X.S.; Maleki, L. Optoelectronic microwave oscillator. *JOSA B* **1996**, *13*, 1725–1735.
4. Chi, Y.-C.; Lin, G.-R. A self-started laser diode pulsation based synthesizer-free optical return-to-zero on-off-keying data generator. *IEEE Trans. Microwave Theory Tech.* **2010**, *58*, 2292–2298.
5. Ghelfi, P.; Laghezza, F.; Scotti, F.; Serafino, G.; Capria, A.; Pinna, S.; Onori, D.; Porzi, C.; Scaffardi, M.; Malacarne, A.; *et al.* A fully photonics-based coherent radar system. *Nature* **2014**, *507*, 341–345.
6. Smit, M.K.; Leijtens, X.; Bente, E.; Van der Tol, J.; Ambrosius, H.; Robbins, D.; Wale, M.J.; Grote, N.; Schell, M. A generic foundry model for InP-based photonic ICs. In Proceedings Optical Fiber Communication Conference OM3E-3, Los Angeles, LA, USA, 4–8 March 2012.
7. Marpaung, D.; Roeloffzen, C.; Heideman, R.; Leinse, A.; Sales, S.; Capmany, J. Integrated microwave photonics. *Laser Photonics Rev.* **2013**, *7*, 506–538.
8. Heck, M.J.R.; Bauters, J.F.; Davenport, M.L.; Spencer, D.T.; Bowers, J.E. Ultra-low loss waveguide platform and its integration with silicon photonics. *Laser Photonics Rev.* **2014**, *8*, 667–686.
9. Tkach, R.W.; Chraplyvy, A.R. Regimes of feedback effects in 1.5 μm distributed feedback lasers. *J. Lightw. Technol.* **1986**, *4*, 1655–1661.
10. Barbarin, Y.; Bente, E.A.J.M.; Heck, M.J.R.; Den Besten, J.H.; Guidi, G.; Oei, Y.S.; Binsma, J.J.M.; Smit, M.K. Realization and modeling of a 27-GHz integrated passively mode-locked ring laser. *IEEE Photonics Technol. Lett.* **2005**, *17*, 2277–2279.

11. Mizumoto, T.; Takei, R.; Shoji, Y. Waveguide optical isolators for integrated optics. *IEEE J. Quant. Elect.***2012**, *48*, 252–260.
12. Galland, C.; Ding, R.; Harris, N.C.; Baehr-Jones, T.; Hochberg, M. Broadband on-chip optical non-reciprocity using phase modulators. *Optics Express* **2013**, *21*, 14500–14511.
13. Tzuang, L.D.; Fang, K.; Nussenzveig, P.; Fan, S.; Lipson, M. Non-reciprocal phase shift induced by an effective magnetic flux for light. *Nat. Photonics* **2014**, *8*, 701–705.
14. Li, E.; Eggleton, B.J.; Fang, K.; Fan, S. Photonic Aharonov-Bohm effect in photon-phonon interactions. *Nat. Commun.* **2014**, *5*, 1–5.
15. Heck, M.J.R.; Bente, E.A.J.M.; Barbarin, Y.; Lenstra, D.; Smit, M.K. Monolithic Semiconductor Waveguide Device Concept for Picosecond Pulse Amplification, Isolation and Spectral Shaping. *IEEE J. Quant. Elect.***2007**, *43*, 910–922.
16. Heck, M.J.R.; Bente, E.A.J.M.; Barbarin, Y.; Fryda, A.; Jung, H.-D.; Oei, Y.S.; Nötzel, R.; Lenstra, D.; Smit, M.K. Characterization of a monolithic concatenated SOA/SA waveguide device for picosecond pulse amplification and shaping. *IEEE J. Quant. Elect.* **2008**, *44*, 360–369.
17. Jones, D.J.; Zhang, L.M.; Carroll, J.E.; Marcenac, D.D. Dynamics of monolithic passively mode-locked semiconductor lasers. *IEEE J. Quant. Elect.* **1996**, *31*, 1051–1058.
18. Bandelow, U.; Radziunas, M.; Vladimirov, A.; Hüttl, B.; Kaiser, R. 40 GHz mode-locked semiconductor lasers: Theory, simulations and experiment. *Opt. Quant. Elect.* **2006**, *38*, 495–512.
19. Heck, M.J.R.; Bauters, J.F.; Davenport, M.L.; Doylend, J.K.; Jain, S.; Kurczveil, G.; Srinivasan, S.; Tang, Y.; Bowers, J.E. Hybrid silicon photonic integrated circuit technology. *IEEE J. Sel. Top. Quant. Elect.* **2013**, *19.*
20. Kurczveil, G.; Pintus, P.; Heck, M.J.R.; Peters, J.D.; Bowers, J.E. Characterization of insertion loss and back reflection in passive hybrid silicon tapers. *IEEE Photonics J.* **2013**, *5*.
21. Barbarin, Y.; Bente, E.A.J.M.; Heck, M.J.R.; Oei, Y.S.; Nötzel, R.; Smit, M.K. Characterization of a 15 GHz integrated bulk InGaAsP passively modelocked ring laser at 1.53 μm. *Opt. Express* **2006**, *14*, 9716–9727.
22. Bente, E.A.J.M.; Barbarin, Y.; Heck, M.J.R.; Smit, M.K. Modeling of integrated extended cavity InP/InGaAsP semiconductor modelocked ring lasers. *Opt. Quant. Elect.* **2008**, *40*, 131–148.

23. Tahvili, M.S.; Barbarin, Y.; Leijtens, X.J.M.; De Vries, T.; Smalbrugge, E.; Bolk, J.; Ambrosius, H.P.M.M.; Smit, M.K.; Bente, E.A.J.M. Directional control of optical power in integrated InP/InGaAsP extended cavity mode-locked ring lasers. *Opt. Lett.* **2011**, *36*, 2462–2464.
24. Heck, M.J.R. Grating Coupler Enabled Optical Isolators and Circulators for Photonic Integrated Circuits.*IEEE J. Sel. Top. Quant. Elect.* **2015**, *21*, 1–9.
25. Tang, Y.; Chen, H.-W.; Jain, S.; Peters, J.D.; Westergren, U.; Bowers, J.E. 50 Gb/s hybrid silicon traveling-wave electroabsorption modulator. *Opt. Express* **2011**, *19*, 5811–5816.
26. Fish, G. Heterogeneous photonic integration for microwave photonic applications. In Proceedings Optical Fiber Communication Conference and Exposition and the National Fiber Optic Engineers Conference (OFC/NFOEC), Anaheim, CA, USA, 17–21 March 2013.
27. Dupuis, N.; Lee, B.G.; Proesel, J.E.; Rylyakov, A.; Rimolo-Donadio, R.; Baks, C.W.; Ardey, A.; Schow, C.L.; Ramaswamy, A.; Roth, J.E.; *et al.* 30-Gb/s Optical Link Combining Heterogeneously Integrated III-V/Si Photonics With 32-nm CMOS Circuits. *J. Lightw. Technol.* **2015**, *33*, 657–662.
28. Thomson, D.J.; Porte, H.; Goll, B.; Knoll, D.; Lischke, S.; Gardes, F.Y.; Hu, Y.; Reed, G.T.; Zimmermann, H.; Zimmermann, L. Silicon carrier depletion modulator with 10 Gbit/s driver realized in high-performance photonic BiCMOS. *Laser Photonics Rev.* **2014**, *8*, 180–187.

Index